FUNDAMENTAL CONCEPTS OF PHYSICS

MICHAEL J. CARDAMONE
Professor of Physics
The Pennsylvania State University

BrownWalker Press
Boca Raton, Florida

Fundamental Concepts of Physics

BrownWalker Press
Boca Raton, Florida
USA • 2007

ISBN: 1-59942-433-9 (paperback)
ISBN 13: 978-1-59942-433-0 (paperback)

ISBN: 1-59942-434-7 (ebook)
ISBN 13: 978-1-59942-434-7 (ebook)

BrownWalker.com

Cover design by Shereen Siddiqui

Library of Congress Cataloging-in-Publication Data

Cardamone, Michael J.
   Fundamental concepts of physics / Michael J. Cardamone.
      p. cm.
   Includes index.
   ISBN 978-1-59942-433-0 (pbk. : alk. paper) -- ISBN 978-1-59942-434-7 (electronic book)
   1. Physics. I. Title.

QC21.3.C37 2007
530--dc22

2007046375

To Barbara, the most important person in my life.

PREFACE

Does the world really need another conceptual physics book? This is a question I asked after teaching the standard one semester course to non-science undergraduates for the past four decades. After examining the available texts, I concluded there was a space for a relatively small, inexpensive text that would present the science of physics in a fashion that did not emphasize the mathematical or computational aspects of the subjects, but instead attempted to introduce some of the reasoning used in the past to arrive at our present understanding of the workings of the physical universe. The following text represents my effort to fulfill this need.

The text does not present the material of physics in the same way we present it to science and engineering students for two important reasons. We train science and engineering students to view physical situations with an analytical eye so that they obtain a quantitative understanding of the situation and are able to make quantitative predictions of observable results arising from the situation. These students think in an analytical mode. Non-science students need not approach physical reality as a series of situations to be analyzed, but more as a mosaic of natural parts forming a whole picture of our world and its place in the universe. In some respects, we expect our technical students to follow the thought processes of an Aristotle while our non-technical students follow those of Plato. Both paths are valid.

You will find very few equations, numerical examples, or problems in this text. Those that are included are for the benefit of the student who appreciates the economy of expressing truths in symbolic mathematical representations and for the instructor who might want to explore some quantitative expectations of the science. The student who is not mathematically inclined need not fear being overwhelmed with numbers.

The included illustrations are simple line drawings by design. While multi-colored artistic representations are useful in many instances, the essence of a concept can often be shown with a minimum of detail. Including only the essential features of a pictorial representation of some physical phenomenon often allows students to see beyond extraneous complications and obtain understanding of the underlying physical situation.

The book is divided into fifteen chapters. Each chapter attempts to introduce large concepts that are applicable to several areas of the physical world. For instance, the concept of momentum and the conservation of momentum in the absence of external forces is central to our understanding of phenomena ranging from the collision of two bodies to the pressure exerted by a confined gas made up of countless particles. For our technical students we would examine these two situations at different times. However, the concept of momentum conservation is central to both situations. For students using this text, understanding of the central concept is more important than the ability to apply the concept to a specific situation where quantitative information is required.

I hope students will approach reading this book more as an intellectual adventure than as an academic chore. When the course is over, if students have a greater understanding of the workings of the universe in which we are all a part and the intellectual developments that lead to this understanding, the book will have been successful.

Good luck in your study of the concepts of physics. I hope you develop a familiarity with the concepts and those who were responsible for their discovery.

TABLE OF CONTENTS

CHAPTER I

PLANETARY MOTION AND UNIVERSAL GRAVITATION

Since the beginning of recorded history, humanity has expressed an interest in the physical world. Individuals attempted to explain phenomena observed in this world on the basis of discovering the causes of the noted effects. The basic assumption was that every effect had a findable cause. This assumption of causality is the basis of our knowledge of the physical universe.

In the earliest recordings of the history of the human endeavor we now call science, individuals had recourse only to superstition and the invocation of supernatural powers. As knowledge and sophistication improved over the course of centuries, supernatural explanations were rejected, as natural laws became known. The object of the study of physical science is to describe observed phenomena in terms of basic, fundamental, universally operative natural laws.

In attempting to attain this objective, the physical scientist reasons in either an inductive (synthetic) or a deductive (analytic) fashion. The former method attempts to first propose universal laws that can be applied to specific problems or phenomena. This is the method often used today in theoretical investigations. The theorist attempts to discover fundamental laws and then tests the validity of these laws by comparing their predictions with the results of experiments or observations. The latter method reverses the steps in that the scientist first experimentally observes a phenomenon or a set of phenomena, then attempts to develop a theory that explains these observations. Both methods are valid and useful in obtaining a general view of nature.

Because both the deductive and the inductive reasoning schemes are valid, it is useful to investigate examples of their application in arriving at our currently accepted ideas on the nature

1

of physical reality. A useful example is the development of the explanation of the problem of planetary motion.

It is likely that the problem of explaining the motion of planets in the night sky is one of the oldest to which humanity gave any attention. Ancient civilizations gave much thought to the problem of the objects wandering through the heavens and arrived at a wide variety of explanations for the observed phenomenon. Some of their explanations seem almost comical today but are, in fact, attempts to explain one of the most fundamental and important mysteries of the ancient world.

The problem of wandering stars received much attention by all the early civilizations, but the explanation stated by the philosopher Ptolemy is perhaps most important because it influenced thinking in Europe for centuries. Ptolemy described the observed universe in terms of a model where Earth stood at the center of a number of spheres of increasing diameter. The moving heavenly bodies, sun, moon, planets, were constrained to move within the spheres on circular paths about Earth. The stars were said to rest on the sphere with the largest diameter and were thought to be the dome of heaven.

Ptolemy's ideas were not seriously challenged until the beginning of the Renaissance. One reason for their long acceptance is that his theory is logically consistent and was not subject to experimental or observational verification until adequately refined techniques and equipment were developed in the fifteenth century. Until that time, logical consistency was sufficient for the acceptance of a theory. The idea that observation and experimentation needed to verify a theory is a relatively recent concept.

Once it became possible to make precise determinations of planetary positions, interest in the problem of planetary motion revived. Tycho Brahe, the Danish astronomer, brought the art of astronomical observation to a zenith in the sixteenth century. Brahe was a firm believer in the Ptolemic system that Copernicus was attacking. He felt accurate observations of planetary motion would verify Ptolemy's idea. With the backing of King Frederick II of Denmark, Tycho built the world's finest observatory where he amassed a huge amount of extremely good data on planetary position as a function of time. Unfortunately, Tycho was a better observer than an interpreter and died before he was able to draw any conclusions from this effort.

FUNDAMENTAL CONCEPTS OF PHYSICS

Johannes Kepler, an assistant of Tycho, obtained possession of Tycho's data trove after his death. With the advantage of having the best observational data ever obtained, Kepler attempted to make the data fit with Ptolemy's concepts. After much effort, he was forced to conclude the theory of Ptolemy had to be abandoned. Only Copernicus' concept of a sun-centered solar system fit the observational data.

In Copernicus' scheme, Earth assumes a position of being only one of several planets traveling around the sun. This displacement of the centrality of Earth in the heavens did not receive ready acceptance from the religious and political authorities in Europe at the time and it was some time before the Copernicus explanation found acceptance. Now, however, we are so sure of its validity that it serves as the basis of our ability to send space probes to other bodies contained within our solar system.

Kepler was an empiricist in that he did not produce any data, but used the data produced by someone else to deduce mathematical relations among the various parameters of interest in the problem. While he did not have a fundamental theory to explain his discoveries, he was highly successful in his endeavor. He was able to compress the data Brahe had complied over a lifetime of observations into three compact laws describing the motion of the planets. Kepler's first law describes the paths of the planets. Simply stated, it is:

The planets move in elliptic orbits with the sun at one focus of the ellipse.

This law is illustrated in the following Figure 1.

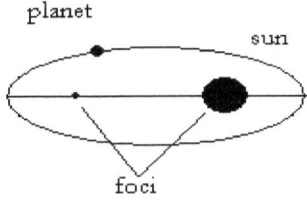

Figure 1.

3

Because the planet moves in an elliptical path, the distance from the sun to the planet is not constant, but varies with time. This conclusion differed from the theory of Copernicus or Galileo who assumed the planets moved in circular paths about the sun. The observational data required a modification to one of the points of the heliocentric theories, but showed those theories to be correct in their essential conclusion that the system was a sun centered rather than an earth centered entity.

Based on his empirical interpretation of the data, Kepler deduced his second law of planetary motion. This law states:

The line drawn from the sun to a planet would sweep out equal areas in equal times.

To illustrate this law consider Figure 2.

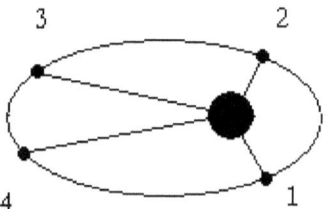

Figure 2.

If the planet is in position 1 at some point in time, after a certain time passes, say for sake of argument two weeks, it is in position 2. After several weeks have passed, the planet is now in position 3. Again after two weeks, the planet has moved to position 4. The line joining the planet to the sun has swept out equal areas when the planet moved from position 1 to 2 as it did when the planet moved from position 3 to 4. Casual observation leads us to the conclusion that the planet was traveling faster when it was nearer the sun than it was when it was farther away. Kepler noted this fact, but needed ten more years of study before he was able to come to a quantified statement concerning it.

Because it is now apparent the motion of planets in their orbits is not one where they travel equal distances in equal times, we need to develop some quantitative understanding of exactly what is meant when we discuss an object's speed or velocity. Although we commonly use the terms speed and velocity interchangeably, the scientist is very specific in differentiating these two concepts. To a scientist, such as Kepler or Galileo, an object's speed is a measure of how fast it moves regardless of direction, while velocity is concerned both with how fast it moves as well as its direction. Quantities such as speed that can be completely specified by a magnitude are named *scalar* quantities, while those needing to be specified by both a magnitude and a direction are called *vector* quantities. There are many examples of both scalar and vector quantities that you will encounter in your study of physical science.

Consider two points, A and B, separated by two meters in an east-west direction. The scalar distance between them is two meters, while the vector displacement from point A to point B is two meters in the west direction. If an object travels from A to B in four seconds, its scalar speed is ½ meter per second, while its vector velocity is ½ meter per second to the west.

As you can see, distance is the scalar magnitude of the displacement vector and speed is the scalar magnitude of the velocity vector. The average velocity to go from A to B is ½ m/s, but that does not mean that the object's speed was ½ m/s at every point on its travel from A to B. It is possible that it took three seconds to go half the distance and another one second to travel the other half. To have a real understanding of the object's motion we need to have a measure not only of its *average* velocity of ½ m/s west, but of its *instantaneous* velocity at every point of its journey. This need led Newton and Leibnitz to develop the mathematical discipline of differential calculus.

Because a vector quantity changes if either its magnitude or direction changes, when we talk about the rate of change of the velocity vector with respect to time that we call *acceleration*, we can focus on the acceleration associated with a change in the direction of the velocity vector, or the acceleration associated with a change in its magnitude. We define *tangential* acceleration as the rate of change of the magnitude of velocity and the *radial* or *centripetal* acceleration as the rate of change of the direction of velocity.

Returning to Kepler's analysis of Brahe's data, we can see the velocity of the planet in its motion about the sun continuously changes in both direction and magnitude. Kepler worked for ten years after he published his first two laws of planetary motion to arrive at some mathematical relationship that would account for the continuous variation of the planets' orbital velocity. He was finally able to relate the time it takes a planet to complete its orbit about the sun, its *period*, to the average distance between the planet and the sun. This he published as his third law of planetary motion: *The ratio of the cube of the distance between planet and sun to the square of the period of its orbit is constant.*

We often find it convenient to write laws such as Kepler's third law in mathematical formulations. If we define the average distance between sun and planet by the letter, r, and the period of its orbit by, T, we can then write this law as:

$$r^3 \propto T^2$$

Example 1. The planet Jupiter takes 12 years to complete one orbit. How far is Jupiter from the sun?

Using Kepler's third law, we can write the ratio of the cube of Jupiter's average distance from the sun to the cube of Earth's average distance equals the ratio of the square of Jupiter's period of revolution to the square of Earth's period.

$$r^3_J/r^3_E = T^2_J/T^2_E$$

Knowing the average distance between Earth and the sun is 93,000,000 miles, we can state the cube of Jupiter's distance from the sun is

$$r^3_J = 12^2/1^2 \ (\ 93{,}000{,}000 \ \text{miles})^3.$$

Or, Jupiter's average distance from the sun is about 487,000,000 miles.

Kepler's three empirical laws of planetary motion represent his unique contribution to our knowledge. It is a tribute to Kepler's energy and genius that the laws of planetary motion carry his name. Remember, Kepler discovered his laws by empirical study. He had masses of very good observational data available to him from which he could glean mathematical relationships. He did not, however, have any fundamental ideas of the forces in nature involved in causing the planets to move in the paths he so completely described.

FUNDAMENTAL CONCEPTS OF PHYSICS

The study of planetary motion was not completed by Kepeler's publication of his three laws.

The next great advance in developing our understanding of planetary motion came from Sir Isaac Newton. Newton studied the motion of objects in order to discover the causes of motion. His approach, the description of an object's motion based on the knowledge of the causes of changes in motion, is called *dynamics*. If the motion is described without regard to the causes of change, the study is called *kinematics*. Kepler described the kinematics of planetary motion. Newton described the dynamics of planetary motion.

Newton was neither an observer as was Brahe, or an empiricist as was Kepler. Instead, he was a theorist. His goal was to put the study of the motion of the planets on a firm theoretical basis. In order to accomplish this, he needed to develop certain mathematical tools and techniques and discover laws that were universally applicable to the motion of material bodies.

Until Newton's time, mathematic relationships were static and expressed truths between variables that held for all time. Newton, realizing that variables could change with time, developed the differential calculus to describe the rate of change of one variable with respect to another.

To illustrate Newton's thinking, consider again our discussion of the velocity of a particle as it moves from point A to point B as shown in Figure 3.

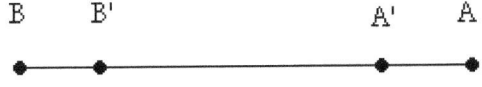

Figure 3.

The magnitude of the average velocity for the particle to move from point A to point B is the ratio of the distance between

7

the two points to the time it takes to go from one point to the other. Now consider the velocity to move from point A' to point B'. The distance between the points has changed and the time needed for a particle to move from one to the other has changed. If, however, the ratio of distance to time has not changed, the average velocity of the particle remains constant. In general, that ratio is different in the first case than in the second. In fact, average velocity can continuously differ along a path leading from one point to another. In order to overcome this difficulty, Newton argued that at every point of an object's journey, a particle had a unique, determinable *instantaneous* velocity.

As we have already seen, the velocity of a planet in its orbit about the sun is not constant but changes with time. Therefore, any completely rigorous mathematical treatment of the problem of planetary motion must, of necessity, use the calculus as developed by Newton.

However, Newton realized that merely having the mathematical tools to describe planetary motion was not enough. He needed to possess an understanding of the fundamental laws of motion of any body. Fortunately, some of these fundamental laws were already known.

Perhaps the most fundamental law concerning the motion of a body is the law of inertia. *A body's inertia is its resistance to a change in its motion.* Newton knew the concept of inertial that had been developed over time and was succinctly stated by Galileo. Before Galileo, it was generally thought that a continuous force was needed to maintain a body's motion. Galileo pointed out that in the absence of any external force acting on a body, its motion would remain constant. Newton appreciated Galileo's thought and stated it in his *First Law of Motion: A object at rest will remain at rest and an object in motion will remain in motion in a straight line with constant speed unless acted upon by an external force.*

Based on this statement of his First Law, Newton realized the planets were under the influence of some external force. To describe planetary motion, Newton realized he needed to answer two questions. How does force acting on a body affect the motion of a body? What is the nature of the force between the sun and a planet?

He answered the first question with his *Second Law of Motion: When acted upon by an external force a body is accelerated in the direction of the force.* Let's consider the implications of this statement. If I take a block of wood and let it drop, is a force acting on it? By Newton's

Second Law, we must say there is a force acting vertically downward that changes the block's velocity from zero when I first release it to some increased value when it hits the floor. If I slide that same block of wood along a tabletop does its motion change? Certainly it starts with an initial velocity but stops after a short time. A force must have acted to slow it to a stop.

Newton needed to arrive at a quantitative understanding of the exact nature of the acceleration imposed on an object by the application of an external force. We can attempt to quantify our ideas by considering the pushing of the block across the tabletop. If we lay down a layer of oil before we push the block, we find the block tends to remain in motion much longer than when the tabletop was dry. If we go farther and drill small holes in the tabletop through which we blow compressed air on which the block floats we find the block once set in motion tends to continue in motion with undiminished speed for a very long time. Let's apply varying forces to the block described in the last case and observe its acceleration. We note when we double the force we double the acceleration as shown in Figure 4.

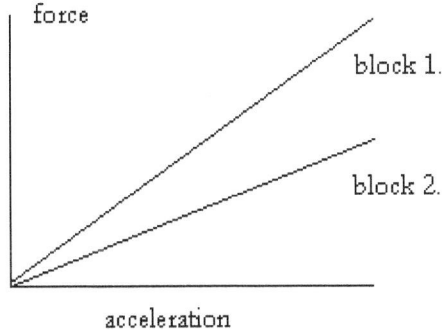

Figure 4.

If we take a second block of the same wood that has twice the volume as the first block, we note the acceleration produced by a given amount of force is just half its value in the first case. We conclude, therefore, there is some intrinsic property of material that represents the relationship between the acceleration an object acquires and the force that produces it. This property of material is called its *mass*. Mass is a measure of the amount of material a body possesses or is a measure of the body's inertia. The greater an

9

object's mass, the greater force must be applied to cause a given acceleration. In a mathematical statement we can write,

$$F = ma$$ Equation 1.

This equation is the quantitative statement of Newton's Second Law of Motion. To be useful, however, we need to define a standard of mass to which any object can be compared. Such a standard is the mass of a platinum-iridium cylinder maintained by the International Bureau of Weights and Measures in France. The mass of this cylinder is defined to be one *kilogram*. With this definition, we can state that a mass of one kilogram will experience an acceleration of one meter per second squared when acted upon by a certain force. This amount of force is called a newton in honor of Sir Isaac.

Example 2. What is the ratio of the acceleration given to an object containing two kilograms of mass to that given to a ten kilogram object when both are subjected to the same force?

Because the force acting on the two objects is the same, the ratio of their accelerations will be inversely proportional to the ratio of their masses

$$a_1/a_2 = m_2/m_1.$$

Therefore, the ratio of the acceleration of the less massive object to that of the more massive object will be 10/2. The less massive object will accelerate five times faster than the more massive one.

In the same manner that we defined a standard unit of mass, we define standard units of length and time. Until 1960, the standard unit of length, the *meter*, was defined as the distance between two scratches on a platinum-iridium bar. Since that time, we defined a much more accurate atomic standard. Likewise, the standard unit of time, the *second*, was defined until 1967 as a certain fraction of a mean solar day. We have also replaced this definition with a more accurate atomic standard.

Concerning the force on the planets causing them to describe paths about the sun, Newton concluded that the force of the sun on the planet must be equal and opposite to the force of the planet on the sun. Generalizing this line of thinking, Newton stated his *Third Law of Motion: For every active force there is an equal and opposite reactive force.*

The force of attraction between the sun and the planets is the same type of force that exists between the earth and the moon, or, in fact, between every pair of massive objects in the universe. This universal force is the force of *gravity.*

Newton, having defined the basic laws concerning the action of forces on objects that we now call *mechanics*, was in a position to investigate the particular nature of the gravitational force required to ensure planets moved in the elliptical orbits described by Kepler. He concluded there exists the *Law of Universal Gravitation.* Simply stated, this law maintains there is an attractive force between every pair of massive objects in the universe that varies as the product of the masses involved and inversely as the square of the distance between them. Written in mathematical shorthand, this is:

$$F \propto m_1 m_2 / r^2.$$

If we introduce a constant of proportionality, we can write this law in an equation:

$$F = G m_1 m_2 / r^2. \qquad \text{Equation 2.}$$

We use the letter, G, as the symbol for the universal gravitational constant. In the system of units we have been discussing, the SI system, the value of G is exceedingly small.

Newton published his law in 1684, but it was more than a century later that Henry Cavendish performed the experiment that determined the numerical value of G. Cavendish published the results of his experiment in 1797. The currently accepted value of the universal gravitational constant is:

$$G = 6.67 \times 10^{-11} \text{Nm}^2 / \text{kg}^2.$$

This number is one of the most fundamental constants in the universe.

You may have heard the romantic story that Newton saw an apple fall and was led to developing his theory of universal gravitation. This story is probably much too romantic to be true. In reality, Newton developed his theory by observing the motion of the moon about the earth and by applying general laws of motion used in conjunction with the mathematical tool of the calculus that he invented. With this analysis, Newton was able to derive Kepler's laws of planetary motion from universally applicable fundamental laws of nature. This is the essence of science, the discovery of natural laws that can account for the ways things work.

Newton's universal gravitation theory stood the test of time for more than two hundred years. It was not until the first part of the twentieth century that his law had to be modified to account for observed results that differed from theoretical prediction. The amount of difference between predictions made using Newton's formulation and the observed results were exceedingly small but were larger than observational error.

Albert Einstein in his *General Theory of Relativity* published in 1915, produced the next great advance in our understanding of gravity. Einstein's theory links the concept of gravity to geometry and will be discussed later in this book. For now, we can say that the predictions of Einstein's theory agree with observations, and we believe the problem of planetary motion is completely solved.

We should now ask ourselves what we have learned from our discussion of the problem of planetary motion. On a shallow level, we learned how one problem developed from observation, to empirical rules, and finally to a theoretical understanding based on fundamental universal physical laws. As we go on we will see examples of other problems that have been solved in the reverse manner, i.e. a theoretical law used to make predictions that are verified by observation or experiment.

On a deeper level, we discovered the force of gravity with which we are intimately associated in our daily lives. We have seen how this force influences the motion of heavenly bodies. In the next chapter, we will investigate gravity closer to ourselves.

QUESTIONS

1. What was Tycho Brahe's contribution to the study of planetary motion? Kepler's? Newton's?
2. What are Kepler's three laws of planetary motion?
3. How does a vector quantity differ from a scalar quantity?
4. What do we mean by velocity? acceleration?
5. How does a centripetal acceleration differ from a tangential acceleration?
6. What is the difference between an instantaneous and an average velocity?
7. What do we mean by "inertia"?
8. What is a body's mass?
9. How does the calculus differ from ordinary algebra?
10. What are Newton' three laws of motion?
11. What was Henry Cavendish's contribution to our understanding of the force of gravity?
12. State Newton's Law of Universal Gravitation.
13. How would things differ if G, the universal gravitation constant, were not such a small number?
14. What contribution did Einstein make to the concept of gravity?
15. Why is it necessary to define standards of length, mass, and time?
16. The United States uses the British system of measurements. What is the unit of length in the British system?
17. List some quantities that are scalar i.e. are completely specified by a magnitude.
18. List some quantities that are vector in nature.

CHAPTER II

QUANTIFICATION AND MECHANICS

In the previous chapter, we were able to solve a specific problem concerning the motion of massive bodies by the application of certain general principles. If these principles really are general, we should be able to apply them to all problems where we attempt to describe the motion of massive objects. This is the goal of *mechanics*.

As with the rest of physics, mechanics seeks to describe phenomena in quantitative terms. In order to achieve this goal, we need to answer two fundamental questions:

1.   What are the essential quantities to be measured?
2.   What units will be used to measure these quantities?

In the study of mechanics, the essential quantities are length, mass and time and the units we use are defined by comparison with the standards presented in the previous chapter. The standard of length is the meter, for mass it is the kilogram, and for time the second. All other quantities that concern us are derived from these fundamentals. Velocity is simply a measure of the ratio of distance to time, acceleration is the ratio of distance to time squared. Force is the product of mass and the ratio of distance to time squared.

The units we use to measure fundamental quantities are somewhat arbitrary. We generally agree to use the SI system of units discussed previously. However, some older texts use a system of units where the basic unit of length is the centimeter, mass is measured in grams, and time in seconds. This system of units is often referred to as the *cgs* system. Because the cgs system was in general use before we standardized on the SI system, many quantities are still commonly measured in these units. For instance, fluid volume is often measured in cc's or cubic centimeters and magnetic fields are often specified in gauss. This is particularly true in the medical profession.

FUNDAMENTAL CONCEPTS OF PHYSICS

Another system of units commonly used in the United States is the British system. In this system, weight, which is the force of gravity acting on a mass, is used. In this system, length is measured in inches, feet, yards, rods, chains, miles, and furlongs. Weight is defined in ounces, pounds, and tons. While time is measured in seconds, minutes, hours, days, fortnights, months, and years. In the British system, a good unit for velocity would be furlongs per fortnight, although this is rarely used. We commonly measure amounts of floor covering in square yards, the amount of gasoline we purchase in gallons, and the amount of heat generated by a propane grill in BTU (British Thermal Units) per hour.

In general, we will use the SI system in our discussion of mechanics. However, from time to time, when it is convenient, we may use either cgs or British units.

Consider the relationship between an applied force and the resultant acceleration it produces on a massive object. In Figure 1, the force will be measured in newtons and the acceleration in m/s^2.

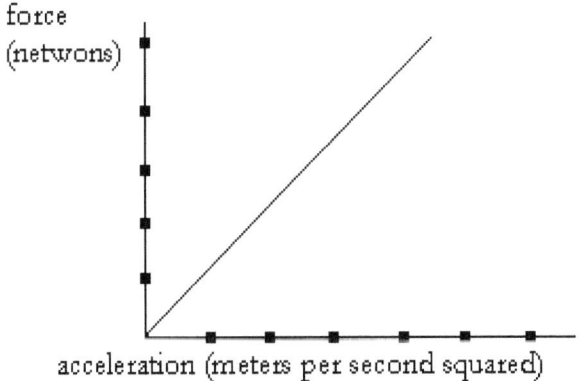

force
(netwons)

acceleration (meters per second squared)

Figure 1.

The newton is the unit of force equal to the product of one kilogram and one meter per second squared. Thus, we have quantified our concept of force to indicate that one newton of force acting on a one kilogram object will impart an acceleration of one meter per second squared. We are now able to ask questions such as the following. What force is needed to accelerate a mass of four

15

kilograms by three meters per second? Or, what acceleration will a six kilogram mass receive when acted upon by a ten kilogram force?

In Figure 1, each tick on the horizontal axis represents an additional acceleration of 2 m/s², and each vertical tick represents a force of an additional 5N. Therefore, we can read off the graph a force or 10N produces an acceleration of 4m/s². Thus, we can conclude the mass being accelerated is 2.5 kg.

You might question the reason for bothering with this level of quantification when we stated we wished to discuss the concepts of physics, not its mathematical computations. The answer is twofold. In the first place, physics itself is a quantitative study. The concepts that are involved are those concepts that lead to quantitative, numerically expressible predictions on the behavior of the matter under study. The heart of the study of physics is often expressed by a quotation of Lord Kelvin that maintains quantitative knowledge is the only legitimate knowledge. I have chosen not to include Kelvin's original statement only because its Victorian language makes it sound pompous and smug. The second reason for introducing this level of quantification is to facilitate the understanding of those ideas that can best be expressed in a quantitative fashion. Often new ideas can be gleaned from old information by investigating the relationships that exist among the quantified variables. Of course, this is the direction taken by Kepler, whose success we noted in the previous chapter.

You may have noticed we used two methods to express the relationship between variables. We either wrote an algebraic equation relating the variables or we drew a graph showing that relationship. When both methods are used to explain a phenomenon, they must contain the same information. This fact was apparent to the French mathematician and philosopher Rene' Descartes, who combined geometric (pictorial) representations with algebraic (numerical) equations to create the study we now call analytic geometry.

Consider again Figure 1 and compare it to the equation representing Newton's Second Law of Motion:

$$\mathbf{F} = \mathbf{ma} \qquad \text{Equation 1.}$$

Figure 1 and Equation 1 contain the same information. The relationship between the applied force and the acceleration it produces is expressed in the equation by the constant, m, while in the graph by the slope of the line i.e. the ratio of the vertical rise of the straight line to its horizontal run.

We can generalize our reasoning to any set of phenomena involving two quantities where one quantity is equal to the product of the other quantity and some constant factor. We call such relationships linear because a plot of one variable vs. the other will produce a straight line. This is true for studies other than those in the physical sciences. Whenever there is a linear relationship between pairs of variables, a change in one will produce a directly proportional change in the other.

We are fortunate if we find a linear relation between variables, but this is often not the situation. Between two variables, say x and y, we may find a relationship that is best represented by a curve as shown in Figure 2.

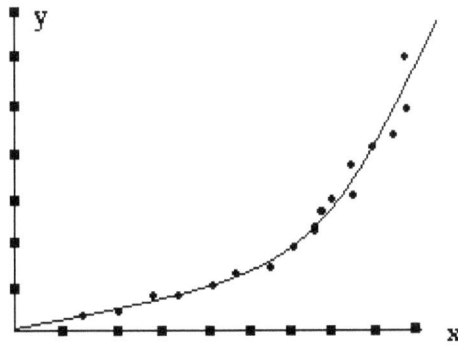

Figure 2

In this situation, the line on the graph best shows the relationship between the variable x and the variable y. We can determine an equation that represents this line, but it will not be a simple statement that y equals the product of a constant factor and x. At best, we may find an equation something like:

$$y = ax + bx^2 \qquad\qquad \text{Equation 2.}$$

In this equation, the coefficients of x and x^2 represent constants.

Obviously, we are not able to discuss the slope of a line that is not linear. However, we can look at two points on such a line and connect them with a straight line. The slope of this line is easily determined. Mathematically, such a line is called a *secant* line. If we allow our two points on the curve to come closer together until they become one point, the slope of the line between the points becomes the slope of a *tangent* line, or stated more precisely, the slope of a line tangent to a curve at a specific point. This is the instantaneous rate of change of one variable with respect to the other that Newton described when he defined the calculus. Applying Newton's technique to Equation 2, we can obtain the rate of change of the variable y with respect to the variable x. This is called the *derivative* of y with respect to x and for Equation 2 is written:

$$dy/dx = a + 2bx \qquad \text{Equation 3.}$$

In the study of mechanics, we often observe the relationship between a variable and time. For example, velocity is the rate of change of position with time, acceleration is the rate of change of velocity with time, and power is the rate of change of energy with time. Time is the *independent variable*. The value of position, velocity, or energy depends on the value of time and so these variables are *dependent variables*.

Example 1. If the position of a body at any time is given by the expression, $x = 6t^2 + 4t$, what is the body's position and velocity after 3 seconds have elapsed?

The position is obtained by substituting $t = 3$ in the above equation to find $x = 6*9 + 4*3 = 66$ units of length. The velocity is given by the derivative of the above expression, $v = 12t + 4$. At a time of three seconds, $v = 12*3 + 4 = 40$ units of length per second.

Let's now apply the quantitative knowledge we have to the problem of a falling body. From our discussion of universal gravitation in the last chapter, we know that an object dropped from some height near the surface of the earth will be attracted to the earth and the earth will be attracted to the falling body. Because the earth is so huge and is so massive compared to any object dropped near its surface, we can observe the motion of any falling body and should be able to deduce laws of motion that would be applicable to any falling body.

About 370 BC, Aristotle proposed an explanation for the behavior of falling bodies based on his concept that all objects were composed of the elements, air, earth, fire, or water. A light object such as a feather contained more air than earth while a stone had more earth than any other element. When a feather and a stone were dropped from the same height, the stone would fall to the ground faster than the feather. Aristotle concluded this was because the stone, containing more earth than the feather, would be attracted to the ground with a greater force and, thus, would reach the ground first. He generalized his conclusion to state that heavier objects would fall faster than lighter objects. This logically consistent explanation of the nature of falling bodies was accepted for almost 2 millennia until Galileo actually studied falling objects and showed in 1589 that different weight objects fall at the same rate.

Galileo introduced the concept of experimental verification that is the basis of our current science. He observed the motion of various weight objects dropped from the Tower of Pisa. He did not accept Aristotle's logically consistent explanation because the results it predicted did not agree with experimental observation. Galileo noted there must be another explanation for the difference in time it took a feather and a stone to fall the same distance. He reasoned the feather's fall was subject to resistance as it fell through the air. If the air were not present, both feather and stone would fall at the same rate. Because he did not have timing devices sufficiently sensitive to make detailed observations of the position of a falling body as a function of time, Galileo devised experiments that used balls rolling down inclined planes.

Today, we are able to make detail measurements of the distance an object falls in a certain time interval. If we consider an object falling through a distance of several meters and observe its position every hundredth of a second, we could plot its position at the end of each time interval to produce a graph shown in Figure 3.

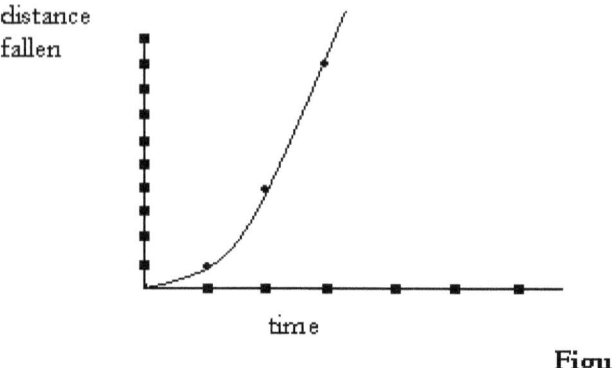

distance
fallen

time

Figure 3.

If we measure the total distance fallen by the object, we note in the first time interval it fell one unit of length. At the end of the second time interval, the total distance fallen is four length units, and after three time intervals it is nine length units. We are tempted to conclude the total distance fallen is proportional to the square of the fall time. Let's produce a graph of distance as a function of time squared, as shown in Figure 4, to test this assumption.

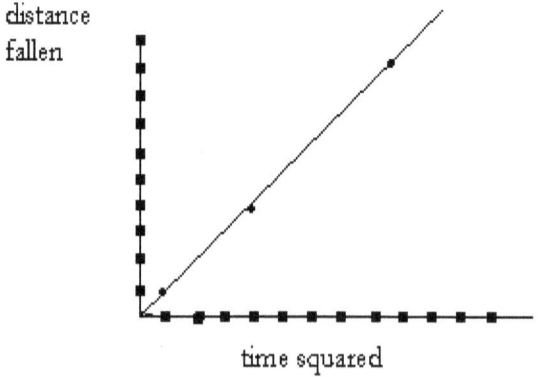

distance
fallen

time squared

Figure 4.

Figure 4 is this plot. It seems, within experimental uncertainty, there is a linear relationship between the distance fallen by a body and the square of its fall time. If we were to throw the object down instead of letting if fall naturally, when we plot the distance fallen against the time of flight squared, we would find a result just like that shown in Figure 4 except with the line displaced vertically upward.

We can get some measure of the velocity of the falling body as a function of time if we compute the average velocity it has during each time interval and use the approximation that the average velocity in any time interval is about equal to the instantaneous velocity at the mid-point of the interval. The shorter the time interval, the closer this approximation is to being exactly correct. When we make these measurements and plot the resulting velocities as a function of time we produce a graph such as that shown in Figure 5.

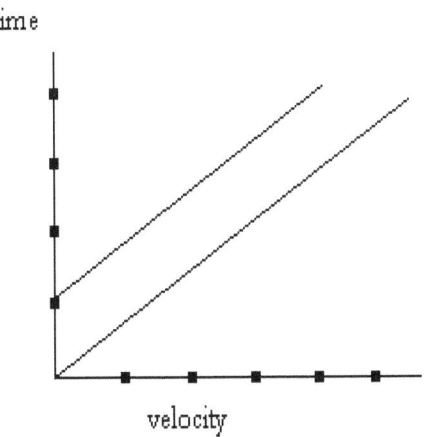

Figure 5.

The lower line in Figure 5 represents the velocity of the object when it is dropped and the upper line represents its velocity when it is thrown down. Note that both lines have the same slope. The slope of the velocity vs. time line is the object's acceleration. From the results shown in Figure 5, we are forced to conclude the acceleration due to gravity of a falling body is constant, regardless of the body's initial velocity. We can succinctly express this fact in the form of an equation.

If y is the distance a body falls, we can write the relation between this distance and the fall time, t, in the following form:

$$y = \tfrac{1}{2}\, gt^2 + v_0t, \qquad \text{Equation 4.}$$

In this equation g is the constant value of acceleration due to gravity, and v_o is the object's initial velocity. If the object were initially thrown upward instead of downward, its initial velocity would be negative.

The object's velocity, the rate of change of its position with respect to time, is given by:

$$v = dy/dt = gt + v_o \qquad \text{Equation 5.}$$

The equations we have obtained investigating a falling body are applicable in any situation where a constant force acts on a body producing a constant acceleration. In the case where the body's acceleration is zero, its velocity is constant and the distance it travels in a certain time is simply the product of its velocity and the time it moves.

When we consider the motion of an object that is moving in two directions, we are able to consider the motion of the object in each direction separately. We then combine the motions to get a general description. An example of this is the motion of a projectile thrown up and out with an initial velocity as shown in Figure 6.

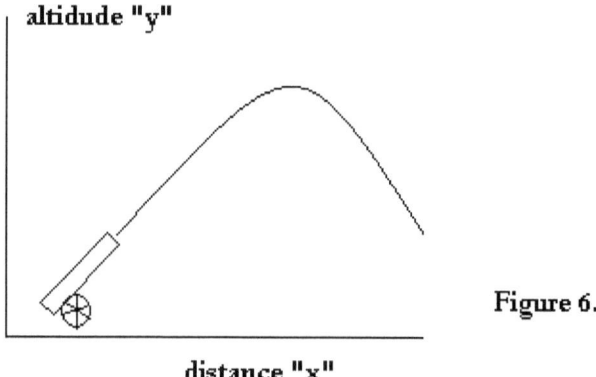

Figure 6.

The projectile has some initial velocity at some angle with respect to the horizontal direction. Because of the vector nature of velocity, we can separate it into *components* that are directed vertically and horizontally. After the projectile is launched, the only force acting on it is the force of gravity that is directly vertically downward.

FUNDAMENTAL CONCEPTS OF PHYSICS

This force produces a constant downward acceleration. In the horizontal direction, there is no force so the horizontal component of the projectile's velocity remains constant throughout its flight. If we define the upward direction as positive, the acceleration due to gravity is in the negative y direction and we can write the projectile's altitude, y, at any time by the equation:

$$y = (v_o)_y t - \tfrac{1}{2} gt^2 \qquad \text{Equation 6.}$$

Likewise, we can determine the projectile's horizontal position, x, at any time by:

$$x = (v_o)_x t \qquad \text{Equation 7.}$$

By eliminating time between the two equations, we are able to write the relationship between the projectile's vertical and horizontal position. Such a description is called the particle's *trajectory*. This equation is written:

$$y = (v_{0y}/v_{0x}) x - \tfrac{1}{2} g x^2 / (v_{0x})^2 \quad \text{Equation 8.}$$

The graph of an equation like Equation 8 is a *parabola*. Therefore, we would expect a real projectile would follow a parabolic trajectory modified to the extent that forces other than gravity affect the object.

QUESTIONS

1. What is the goal of mechanics?
2. What are the fundamental questions that need to be answered in order to quantitatively describe physical phenomena?
3. What system of units is commonly used for scientific measurement?
4. What force is needed to impart an acceleration of $5m/s^2$ to a 10 kg mass?
5. Who was first to combine successfully the concepts of algebra and geometry?
6. What is meant by a linear relationship between variables?
7. What is the difference between average and instantaneous velocity?
8. What is the definition of a derivative?
9. What was Aristotle's explanation for the motion of a falling body?
10. What did Galileo do for the study of freely falling bodies?
11. Why does a stone fall faster than a feather when both fall under the influence of gravity?
12. How is the position of a freely falling body related to its flight time?
13. How does the velocity of a freely falling body vary with time?
14. If an object is thrown vertically upward why will it not continue to rise?
15. Why will a projectile fired horizontally and another freely dropped from the same height strike the ground simultaneously?
16. What is the relationship between a projectile's vertical position and its horizontal position called?

CHAPTER III

FORCES OTHER THAN GRAVITY

Until now, we confined our discussion to the motion of bodies under the action of a force, and we examined in detail the force of gravity that exists between every pair of massive objects in the universe. If this were the only force that existed, our task of explaining the motion of bodies would be very simple. However, this is not the case and there are other natural forces that we must investigate in order to gain an understanding of the workings of the physical universe.

Consider the situation illustrated in Figure 1, where a small pith or cork sphere is hanging by a thread. We know the small sphere is attracted downward to the earth by the force of gravity. Now, bring a hard rubber rod near the sphere. The rod and the sphere are attracted to the other by the gravitational force. Because the sphere and the rod contain such little mass, we are unable to easily see the attractive force between them and see only the force of the earth on the sphere.

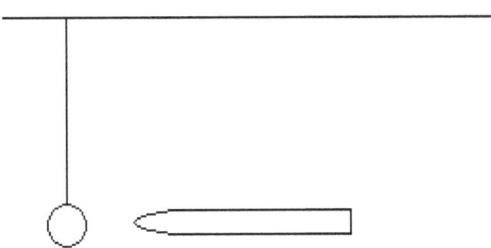

Figure 1.

Let's now rub the hard rubber rod with animal fur and observe what happens. The situation changes as shown in Figure 2.

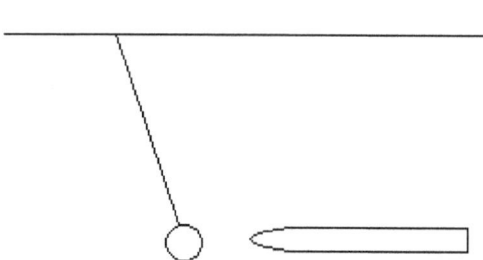

Figure 2.

We notice the string no longer hangs vertically and there is a significant attraction between the sphere and the rod. If we allow the sphere to touch the rod, something even stranger occurs. The sphere stays in contact with the rod for a short time then is repelled from it as shown in Figure 3.

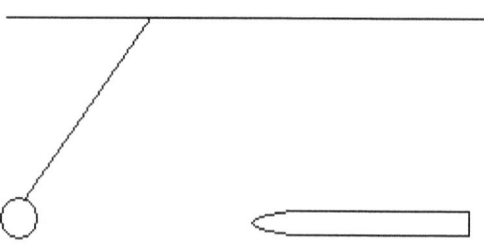

Figure 3.

Obviously, the act of rubbing the rod with the fur has had a significant effect on the way the two objects interact. We can deduce from our experiment that the mere act of rubbing the rubber rod with fur has caused a new type of previously unknown force to come into existence. This force is stronger than the gravitational force and can at times be repulsive as well as attractive. This is something new.

But, is this really something you have never seen before? Perhaps as a child you rubbed a rubber balloon on you hair then stuck it onto a wall. Until you rubbed the balloon in your hair, it would not hold onto the wall and would fall to the floor. Once you rubbed it on your hair, it could stick to the wall. You may have observed this trick was easier to perform on a dry day.

What you did with balloons as children is not significantly different than what we did in observing the interaction of the rubber rod and the pith sphere. Until we rubbed the rod on the fur, we were unable to notice any interaction between it and the sphere. Once the rod was in some way modified by being rubbed with fur, there was a noticeable interaction between it and the sphere.

Let's repeat our observations with a series of pith spheres that we touch with rods of various materials rubbed with other materials. In a specific instance, let's rub a glass rod on silk. When we bring this rod near a pith sphere, we observe the same type of interaction we previous saw with the hard rubber rod.

If we produce a number of pith spheres that were touched with the rubber rod and another number that were touched with the glass rod, we can investigate the interactions between these two types of spheres. When we bring two spheres that had touched the rubber rod near each other, we observe there is a strong repulsion between them. When we do the same thing with two spheres touched by the glass rod, we notice the same effect. However, when we bring together one sphere touched by the rubber and one touched by the glass, we notice a strong attraction between them. After numerous observations, we conclude there exists a repulsive force between objects that have been changed by being touched to the same rod and an attractive force between objects touched by different rods. We also conclude, based on repeated experiments with multiple materials, that there are only two types of modifications that the sphere may attain.

In the mid eighteenth century, these investigations lead to the idea that materials contained *electric charges*. Benjamin Franklin, identified these charges as being one of two types that he called *positive* or *negative*. Before a rod is rubbed, it contains equal number of positive and negative charges. The rubbing transfers some charge to or from the rod such that the process renders the rod with a net positive or negative charge. On an arbitrary basis, *we define the type of charge a glass rod acquires when rubbed with silk as being positive and the type of charge a rubber rod acquires when rubbed with fur to be negative.* On the basis of this definition and our observations, we can conclude like charges repel each other and opposite charges attract each other.

But, how does this explain the initial attraction between the electrically charged rod and the uncharged pith sphere? We can explain this if the electric charges in the sphere can reorient themselves within the sphere *and* if the force between charges decreases with the distance between the charges. We know gravitational force decreases with distance so it is not surprising that electrical forces may decrease with distance. If charges can transfer to and from charging rods they can reorient in pith.

We have arrived at a fairly good qualitative understanding of the force that exists between objects that are electrically charged. What we now need to do is extend this to a *quantitative* understanding. This quantitative study of charged objects where the charges are stationary is called *electrostatics* as opposed to *electrodynamics* that deals with the interactions that occur when charges are in motion.

Until now, everything we measured was in terms of mass, length, and time. In ordered to quantify our understanding of electrostatics and electrodynamics, we need to define a fundamental quantity of charge. This fundamental quantity is called the *coulomb* in honor of Charles Augustin de Coulomb who quantified the law of forces between charge bodies.

Coulomb used a torsion balance to measure the forces between two charged objects. This balance consisted of a torsion fiber with a bar attached to its end as shown in Figure 4.

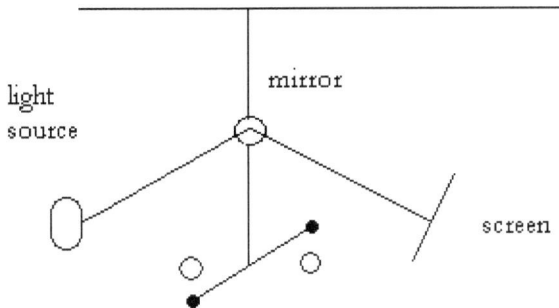

Figure 4.

A small mirror was attached to the fiber and light metal spheres were attached to the ends of the bar. The balance fiber was vertical and other metal spheres were placed horizontally in the plane of the bar. If a force existed between the spheres on the bar and those in the horizontal plane, the fiber would twist. A light shined onto the mirror would move across a screen as the mirror twisted with the fiber. By measuring the amount of twist, Coulomb was able to obtain a measure of the force between the spheres. While not being able to determine the absolute amount of charge he could place on the spheres, Coulomb was able to make successive observations with proportionate amounts so as to deduce the relationship between the amount of charge on the spheres and the magnitude of force. Additionally, he could adjust the distance between the spheres on the bar and those in the plane to determine the nature of the force's distance variation.

Coulomb published the results of his experiment in 1785. He determined *there exists a force between charged bodies that varies directly as the product of the charges and inversely as the square of the distance between them.* We now call this statement Coulomb's Law. Symbolically, if we use, q, to indicate the magnitude of charge on an object, we can write an equation that expresses Coulomb's Law.

$$F = kq_1q_2/r^2 \qquad \text{Equation 1.}$$

If you remember our algebraic statement of Newton's gravitational law, you will recognize that both laws look very similar, with the force between objects proportional to the product of some property common to the bodies and inversely proportional to the square of the distance between them. In general, we classify both as "inverse square" laws. The difference between the laws, however, is very significant. In the first place, the force of gravity is always attractive while the electrostatic force may be attractive or repulsive. Additionally, in 1797 Henry Cavendish used a torsion balance similar to Coulomb's to measure the gravitational proportionality constant. In modern SI units the measured value or this constant is

$$G = 6.67 \times 10^{-11} \text{ N m}^2 / \text{kg}^2.$$

The electrostatic force constant, k, is significantly different. In SI units, this constant is

$$k = 9.0 \times 10^9 \text{ N m}^2 / \text{C}^2.$$

The numerical difference of these constants is enormous. It means if we had two objects, both with 1 kg mass and an excess charge of 1 C, the electrostatic force would be greater than the gravitational force by some twenty orders of magnitude, a truly huge difference.

Beyond electrostatic forces, we are familiar with another force that can be either attractive or repulsive. This is the magnetic force that was known by Greeks and Chinese since antiquity. They knew some materials existed in nature that possessed the ability to attract pieces of iron. Both civilizations discovered that natural magnetic material, such as loadstone, allowed to rotate freely in space would align in a north-south direction. This knowledge was immensely practical in allowing the development of commerce across seas. Although the ancients did not understand the force that aligned their magnets, they did know it was not the same force that caused an object to fall.

Today we can produce magnets by magnetizing pieces of material such as iron. When we make such magnets in the shape of a bar, we soon discover that any bar magnet allowed to rotate freely will align with one end always pointing north and the other end always pointed south. We identify the ends of the bar as being either north seeking or south seeking. We will identify one pole with an S and the other pole with an N, as shown in Figure 5.

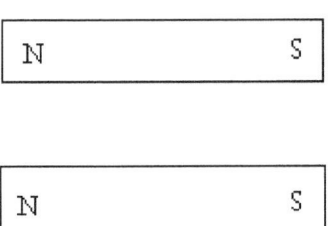

Figure 5.

When we bring these two magnets near each other, we notice they tend to reorient in such a way that the pole marked with an N on one magnet is attracted to the pole marked S on the other magnet. If we attempt to push the two poles marked N or the two pole marked S together, we find they repel each other. Therefore, we conclude *like magnetic poles repel and unlike magnet poles attract each other.*

Let's attempt to isolate the poles of a bar magnet by breaking it as shown in Figure 6.

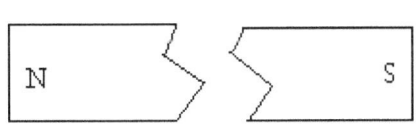

Figure 6.

When we do this, the end marked N develops a south pole at its broken end, and the end marked S develops a north pole at its broken end. Regardless of how small we break the magnet, we are not able to isolate a single magnetic pole in the way we can separate positive electric charge from negative charge. It seems we are not able to produce magnetic monopoles in the same way we can produce isolated electric charges. However, we still want to quantify our understanding of magnetic forces.

As with our investigation of the electrostatic force and the gravitational force, we want to find a quantitative description of the force between two magnets. There are several ways to accomplish this, but we can make use of a torsion balance as did Coulomb and Cavendish. In so doing, we discover the force between magnets varies inversely with the square of the distance between the magnets. If we define some measure of the lifting power of an individual magnet, usually called its *pole strength* signified by, m, we find the force between magnets varies directly as the product of the pole strengths. We again can write a force equation:

$$F = k'm_1m_2 / r^2 \qquad \text{Equation 2.}$$

We may now be anxious to attempt to find a connection between magnetic forces and electrostatic forces. Both forces are much stronger than gravitational forces and both can be either attractive or repulsive. To investigate a possible connection, lets bring a bar magnet near a pith sphere that has been charged. When we do so, nothing happens. Regardless of the conditions we impose, we are unable to detect any interaction between magnets and static charges. Could there be an interaction between magnets and moving charges?

In order to investigate the possible interaction between magnets and moving charges, it is necessary to have some source of moving charges. Fortunately, the investigations of Luigi Galvani and Alessandro Volta in the late eighteenth century developed devices now called *voltaic cells* or *batteries* that used chemical reactions to produce a source where charge would continually flow from one post or *terminal* to another if they are externally connected by a piece of metal. We call materials such as metal where charge can easily flow *conductors*. If the terminals were connected with a piece of wood or rope, there would be little or no flow of charge. Such materials are called *insulators*. The rate of charge passing through a conductor per unit time is called an *electric current*. Originally, we thought this current represented the passage of positive charge carriers in a particular direction. We now know, in fact, it is negative charge carriers moving in the opposite direction that cause the current to exist. In many instances, the exact nature of the charge carriers is irrelevant. We obtain the same net amount of charge passing through an area per unit time if we have positive charge moving in one direction or negative charge moving in then other.

Now consider what happens when a conductor is brought near a magnet. Andre Ampere, in 1820, reported the results of his experiments on the interaction of magnets with current carrying conductors. He showed there was a force between the magnet and the conductor that would differ in direction depending on the direction of the current. This situation is illustrated in Figure 7.

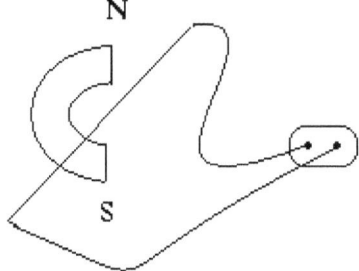

Figure 7.

A wire is placed between the ends of a horseshoe magnet and is attached to a battery. As soon as the connection is made, the wire is pulled toward the magnet or is thrown away form it. If we reverse the terminal connections to reverse the flow of current, we notice just the opposite interaction. Obviously, there is an interactive force between the magnet and the wire that depends on the current in the wire. If we use a stronger battery to provide a greater amount of current, we discover the force is directly proportional to the current in the wire and its direction depends on the direction of the current.

Hans Christian Oersted performed a complementary experiment to Ampere's by noting what happened to a magnetized needle when it was in the vicinity of a current carrying conductor. He observed the needle was deflected due to the current. However, the deflection was not directed towards or away from the conductor but rather was in such a direction that the needle pointed in direction perpendicular to the direction of current flow. When the needle was moved around the wire, it would always point in a direction such that the force on the magnetic needle existed in circular paths about the wire. If the current direction were reversed, the circular paths about the wire would also be reversed.

One other fundamental experiment concerning the interaction of currents and magnets had to await Johann Schweigger's development of an instrument called a *galvanometer*. This instrument had an indicator whose deflection was proportional to the current that ran through it. Michael Faraday in Great Briton and Joseph Henry in the United States used the galvanometer to further investigate the interaction between magnets and electric currents.

If the battery in the experiment shown in Figure 7 is replaced by a galvanometer, nothing happens unless the wire is physically moved. When it is moved, however, a current is detected that is proportional to the speed with which the wire is moved. If the wire is moved away from the magnet, the current is in one direction and if it is moved toward the magnet, the current is in the other direction. The current is *induced* in the wire due to its motion near the magnet. Faraday extensively studied this phenomenon of *electromagnetic induction*, which is the basis of electric generators, motors, and, in a very basic sense, our entire electronically based society.

So, what have we learned from our discussion of forces other than gravity? One thing we learned is gravitational force is not the only force we need to consider. We observed that the force between static electric charges, the force between magnets, and the force between currents and magnets can be attractive, repulsive, or at some direction perpendicular to an object. Also, we found these forces can be much stronger than the gravitational force.

QUESTIONS

1. How is Coulomb's Law similar to Newton's Law of Universal Gravitation?
2. How does Coulomb's Law differ from Newton's Law of Universal Gravitation?
3. How can we explain the initial attraction then the repulsion between a pith sphere and a fur rubbed rubber rod?
4. How do we define positive and negative electric charge?
5. Describe Coulomb's experiment.
6. What is the fundamental unit of electric charge?
7. How does the proportionality constant of Coulomb's Law compare to that of Newton's Law?
8. Why is the magnetic nature of the earth important in the development of commerce?
9. What can we do to investigate the interaction between magnets and static electric charges?
10. How do we differentiate electric conductors from insulators?
11. What do we mean by electric current?
12. How do Ampere's experiment and Oersted's experiment complement each other?
13. What do Henry's and Faraday's experiments indicate?
14. What do we mean by electromagnetic induction?

CHAPTER IV

ENERGY

Until now we have been concerned with the problem of determining the position and motion of an object at some future time if we know its current position, velocity, and the forces acting on it. Formally, we can describe the motion of any object in this way. In a practical sense, however, it is often difficult if not impossible to determine all the forces that can act on an object. In addition, we need not only to specify magnitude of the acting forces, we must specify their directions as well. It would be much easier if we could describe motion by using some *scalar* quantity associated with an object's motion rather than the sum of all forces that are *vector* quantities.

It certainly would be to our advantage if we could find another way to describe the motion of a body without the need to specify the nature of the forces acting on it. Let's see if we can find a method to determine how an object interacts with its surroundings without being explicit about the forces involved.

Consider throwing a small metal ball into a piece of clay as shown in Figure 1.

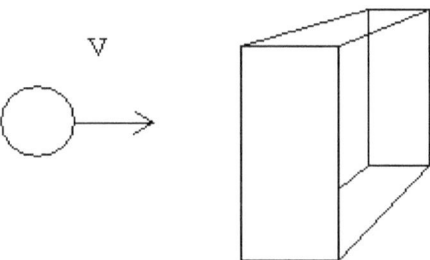

Figure 1.

When the ball hits the clay brick, it leaves a dent or depression. If we throw the ball faster, it leaves a bigger dent. If we repeat this experiment several times, we come to the conclusion that the faster we throw the ball, the greater is the damage to the clay brick.

We can repeat the experiment with a larger, more massive ball. When we do, we find the more massive ball does more damage than does the smaller ball. This should come as no surprise.

Let's see how the damage we have been doing relates to the balls' masses and velocities. We can do this by letting the balls fall from measured heights so they that we know their relative velocities when they hit the clay brick.

From repeated observations, we can conclude the damage done to the clay varies proportional to the mass of the object that collides with it. When we examine the damage done by one ball hitting with varying velocity we conclude the damage is proportional to the *square* of the velocity. We are going to define the falling object's *energy* as its ability to damage the clay brick. We will define the object's *kinetic energy* as the ability it has to cause damage because of its motion. To quantify this definition, we will introduce ½ as the proportionality constant between kinetic energy and the product of mass with velocity squared to write Equation 1.

$$E_k = \tfrac{1}{2}\, mv^2 \qquad\qquad \text{Equation 1.}$$

In the SI units we have used to measure, a mass of two kilograms moving with a speed of one meter per second will have energy of one *joule*. This unit is named for James Joule who was instrumental in the development of the concept of energy in the mid nineteenth century.

Example 1. What is the kinetic energy of a 1 kg mass moving at 25 m/s? Using the quantitative definition in the above equation, the mass possesses a kinetic energy of 312.5 joules.

If we think a bit more deeply about our experiment, we can conclude our falling ball had the potential to do damage to the clay block before we allow it to fall. The simple act of lifting the ball over the brick increases its potential ability to deform the brick. It has energy because of its position as well as its motion. We call such energy *potential energy*. In this specific instance, because the ball has the ability to fall under the influence of gravity, it has *gravitational potential energy*. In other situations, objects can possess potential energy because of their positions at the end of a compressed spring or several other positions. These potential energies will become important in various situations and will be introduced when needed.

Let's consider the total energy an object has when it is raised a certain height above the clay brick we placed on the floor. From our previous study of a freely falling body, we know that the velocity a body acquires when it falls through a distance, h, is given by:

$$v^2 = 2gh \qquad \text{Equation 2.}$$

If both sides of this equation are multiplied by ½ m to produce the expression for the object's kinetic energy on the left hand side of the equality, we can identify the object's gravitational potential energy as the expression on the right hand side.

$$E_p = mgh \qquad \text{Equation 3.}$$

During the time the object falls, the sum of its kinetic energy and its gravitational potential energy remain constant and is equal to the gravitational potential energy it had when initially released. That is, during its fall the object's total mechanical energy is *conserved*. The principle of *conservation of energy* is a central concept that is useful in formulating a new way of examining motion that does not need to identify forces.

Let's now consider sliding a wooden block across a tabletop. We can give this block an initial velocity and note that it will eventually stop. As we observed previously, if we oil the tabletop the block will slide farther before it stops. And, if we drill small holes in the tabletop to allow compressed air to form a cushion between the block and the tabletop the wooden block will travel even farther before it eventually stops. From our knowledge of Newton's laws, we know a force has acted to stop the block. We identify the force that opposes the motion of the block as the force of *friction* between the block and the tabletop.

When we look at this same problem from the basis of the block's energy, we conclude the block initially had some kinetic energy that decreased as it slowed to a stop. Once stopped, it had no mechanical energy as long as it stayed on the tabletop. What happened to our great idea of energy conservation?

The answer is that the block's kinetic energy was dissipated in working against the force of friction. In a situation where friction is present, the total mechanical energy of a system is *not* conserved. Where does the energy go in such a situation?

Here we appeal to an observation you can make on a cold day when you rub your hands together. The friction between your hands tends to make them warmer. The energy you expend in rubbing your hands works against the frictional force opposing their motion and appears as *heat*. This is a clue to us that heat is just another form of energy. While we now know this is true, it required a classic experiment performed by James Joule to verify this fact. Friction is an example of a non-conservative force, i.e. a force where total mechanical energy is not conserved.

Let's now return to a conservative system. Consider a pendulum as shown in Figure 2. As the pendulum moves to the right, it rises through a height, h. That means its gravitational potential energy has been increased by an amount

$$\Delta E_p = mgh \qquad\qquad \text{Equation 4.}$$

The symbol, Δ, simply means the change in the amount of the gravitational potential energy. When the pendulum is released, it will swing through an arc until it reaches a low point. At this time all its acquired potential energy would have converted into kinetic energy and the pendulum would be traveling with its maximum speed. As it rises to the left, the kinetic energy it had at the bottom of its arc will again become potential energy as it rises. When it rises to h, it will have converted all its kinetic energy into potential energy and its speed will be zero. It will then swing in an arc back to the right. This action would continue indefinitely if there were no friction to dissipate the mechanical energy.

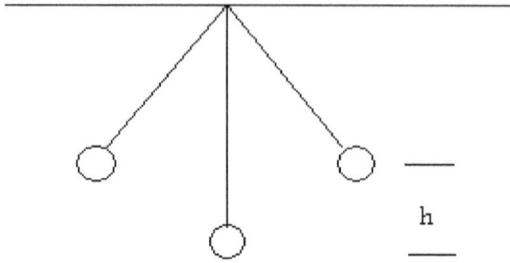

Figure 2.

Very long pendulums are often used to demonstrate the rotation of the earth. The plane of the pendulum's swing remains constant in space while the earth rotates under it. For example, if a pendulum starts swinging in a north-south plane, after a period of six hours it will swing in an east-west plane. After another six hours passes, it will again swing in a north-south plane. After a full day, the pendulum will be swinging in the same direction as it had the day before. Energy conservation worked to keep the pendulum swinging while the earth rotated beneath it. Jean Foucault in 1851 first demonstrated this phenomenon. Since then, it has been again used many times in buildings dedicated to showcasing scientific and technological advancements.

Another system we can consider is that of a mass hanging under the influence of gravity at the end of a stretched spring. If we let the mass undisturbed, the force of gravity acting on the mass, its weight, is balanced by the force of the spring pulling it upwards. We can investigate this latter force by placing differing masses on the spring and noting how the spring responds. When we perform this study, we discover the force the spring exerts on the mass is directly proportional to the amount the spring stretches. If we use, x, to represent the displacement from equilibrium, we express this fact in an equation:

$$F = kx \qquad \text{Equation 5.}$$

When we have the spring-mass system in equilibrium, i.e. the force of gravity on the mass is exactly balanced by the force of the spring on the mass, we can disturb the equilibrium by using some external force to pull the mass down. Upon releasing the mass, the spring compresses and accelerates the mass upward. When the mass passes through its equilibrium position, it does so with some velocity and overshoots the position. The mass does not stop until it rises a height above its equilibrium position that equals the distance below its equilibrium position it was originally pulled down. It seems clear we added some potential energy to the system when we pulled the mass from its equilibrium position. When we released the mass, this potential energy transformed into kinetic energy that had a maximum value when the mass traveled through the equilibrium position. This kinetic energy transformed into potential energy as the spring compressed. At the top of its path, when the mass momentarily stopped, all the mechanical energy was potential. This exchange between kinetic energy and potential is mathematically identical to the situation that existed with the pendulum system if we identify the *elastic potential energy* of the spring-mass system as

$$E_p = \tfrac{1}{2}\,kx^2 \qquad\qquad \text{Equation 6.}$$

The elastic force is a conservative force in the same sense as the gravitational force. It seems there should be a relationship between conservative forces and associated potential energies. Such a relationship does exist when we say that the *work,* defined as *the product of a force and the distance it acts,* done by a conservative force produces an increase in the potential energy of the body on which the force acts.

The pendulum problem and the spring-mass problem are examples of oscillatory motion where there is a continuous exchange of energy between potential and kinetic form. When there is no dissipative force present, we call such motion *simple harmonic motion, SHO.* There are some characteristics that are common in all forms of SHO. The maximum excursion of an object, e.g. the mass on the spring or the pendulum bob, from its equilibrium position is called the *amplitude,* of the motion, the time it takes to complete one oscillation is the system's *period,* and the number of periods per unit time is the motion's *frequency.*

The period of SHO is independent of the amplitude of the motion. Galileo first described this fact in 1581 after he observed the pendulum motion of a lamp in the cathedral of Pisa. He observed the swinging lamp and timed its period using his pulse as a timing device. His observation, made very early in his career, was characteristic of the careful use of observation and deduction that marked his life.

In any system undergoing SHO, we can measure the excursion of some object from an equilibrium position as a function of time. Consider a SHO with amplitude of 10 units and period of 5 seconds. If we plot displacement as a function of time, starting our timing device when the object is at its equilibrium position, we produce a graph as shown in Figure 3.

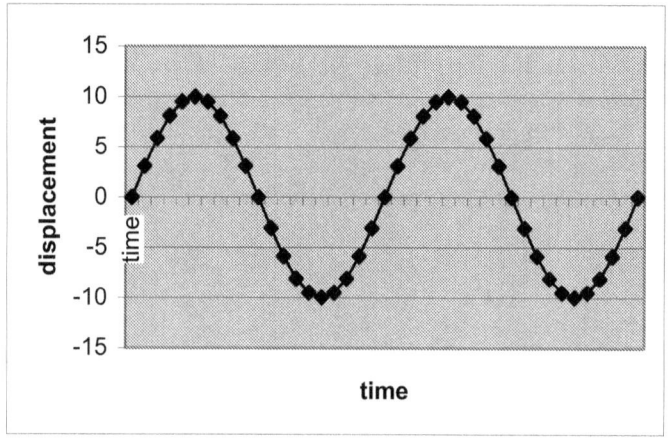

Figure 3.

For the spring-mass system, actual computations are no simpler using energy principles or using Newton's force laws directly. For the pendulum system, however, computations are significantly easier when using energy considerations. The reason is that energy is a *scalar* quantity while force is a *vector*. To determine the net force acting on the pendulum bob at each point in its swing, we need to note that the bob's weight is always directed vertically downward, but the direction of the force of the string on the bob varies as the bob swings. To quantitatively describe the motion, we need to know how these forces act at each instant of time during the pendulum's swing. The situation is considerably easier when we need only to know the pendulum's initial energy and its height above the low point in order to determine the bob's velocity. This use of energy principles greatly simplifies problem solving in complex motions where the direction of force action varies during the course of the motion.

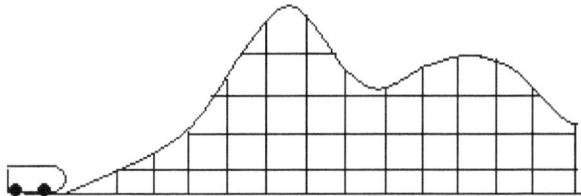

Figure 4.

As an example, consider the problem of the roller coaster illustrated in Figure 4. In this instance, the coaster car is hauled to the top of a high incline, released at the top of the incline, and coasts down the rails until it ends its journey at a position lower than its starting point. During the course of its travels, it goes up and down hills, and around curves, but the car never rises as high as it was when first hauled up the initial hill. If the car is lifted up the first hill and is at rest before it starts its downward run, all of its initial energy is in the form of gravitational potential energy. After it drops a certain distance, some of this initial energy is transformed into kinetic energy and some is still potential energy. The car will continue to possess some mixture of kinetic and potential energy throughout its path, and we can determine its speed at any point on the path as long as we know its height.

The situation we described is idealized in the sense that there are always some frictional forces involved with the motion that we neglected. In a real situation, the coaster car has some small velocity at the top of the first incline so its total energy is the sum of its gravitational potential energy and its initial kinetic energy.

As you can see, the concept of energy is quite powerful, and has much broader applications than we discussed. Ultimately, the concept of energy can be applied in any situation where there is an ability to produce movement. The mechanical energy we have been discussing is just one form of energy. There are many types of energy yet to be discussed. There is energy contained in the random motion of the molecules of a gas, energy stored in the fuel we consume to move our cars and trains, energy transmitted over copper wires into our homes to heat and cool areas, energy in the heart of the atom that can be released over time to power cities or quickly to destroy them. The concept of energy is one of the central concepts of our physical universe. An understanding of energy is essential for gaining knowledge of this universe.

QUESTIONS

1. What is the difference between a scalar and a vector quantity?
2. Give some examples of scalar quantities.
3. Give some examples of vector quantities.
4. What is energy?
5. What is the strict definition of work?
6. How does potential energy differ from kinetic energy?
7. What do we mean by conservation of energy?
8. What do we mean by a conservative force?
9. Give an example of a force that is conservative.
10. What is a non-conservative force?
11. Give an example of a non-conservative force.
12. When mechanical energy is not conserved, where does it go?
13. How does gravitational potential energy vary with height?
14. How does elastic potential energy vary with displacement from equilibrium?
15. What is the amplitude of an object in simple harmonic motion?
16. How are frequency and period in SHO related?
17. Can a roller coaster car ever rise higher than the height of the first incline if it starts from rest at the top? Why?

CHAPTER V

HEAT AND ENERGY

In the last chapter, we considered the warming effect of working against a frictional force. This would lead us to consider there is a connection between energy and heat. In fact, there is a very strong connection and we came to realize in the middle of the nineteenth century that heat was just another form of energy. This concept was proposed by Justin von Mayer in 1842 and was experimentally demonstrated by James Joule one year later.

Joule considered the situation where a container filled with water is heated. If we only observe a thermometer placed in the water, we note the addition of heat results in a rise in temperature. Joule considered what would happen if a similar container of water were stirred by a paddlewheel. If we focused on a thermometer placed in the water, we would see a rise in temperature similar to that which occurred when the water was heated. This observation indicated there was an equivalent effect on the water's temperature if heat flowed into the water or if mechanical work was done stirring the water.

Unfortunately, we had come to the middle of the nineteenth century thinking of heat and energy as different entities and had developed different units for measuring each. Historically, we had to devise some way to measure the relative hotness or coldness of a body. One of the first recorded attempts to do this was performed by Gabriel Fahrenheit in 1709 when he perfected the alcohol thermometer.

Fahrenheit noticed the height of a column of alcohol contained within a capillary tube connected to a reservoir bulb, as shown in Figure 1, rose or fell with the relative hotness or coldness of its surroundings.

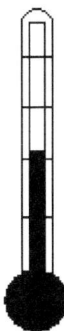

Figure 1.

He needed to establish some starting point for his measurement of temperature and chose freezing seawater as his zero of temperature. To fix an upper point so as to establish a temperature scale, legend has it he chose the temperature of the human body. Perhaps the person he chose to establish this temperature point was running a slight fever. In any event, the 100 degree point was established and the Fahrenheit temperature scale was defined.

The Fahrenheit scale has a great problem in its reproducibility. We know seawater can have various freezing points depending on the amount of salt it contains, and the normal body temperature is not 100 degrees, but a slightly lower figure. Anders Celsius, in 1742, defined a more sensible temperature scale. He chose the freezing point of pure water as his zero point and the boiling point of water under standard atmospheric pressure as his 100 degree point.

On the Fahrenheit scale, water's freezing point is 32 degrees and its normal boiling point is 212 degrees. Therefore, a temperature difference of 180 degrees Fahrenheit is equal to a difference of 100 degrees Celsius. The Celsius degree is about twice as large as the Fahrenheit degree. Unfortunately, neither the Fahrenheit nor the Celsius scale defines an *absolute zero* of temperature. Both are *relative* temperature scales. Later it will become necessary for us to consider an *absolute temperature scale*.

The alcohol in glass device is not the only thermometer we can devise. Any thing with a property that varies with temperature can be used as a thermometer. For example, the electrical resistance of a resistor varies linearly with temperature. Additionally, in a gas contained in a cylinder capped with a movable piston, the volume of the gas varies directly with temperature, and in a constant volume cylinder, the pressure of the contained gas varies directly with temperature. This latter fact is illustrated in Figure 2.

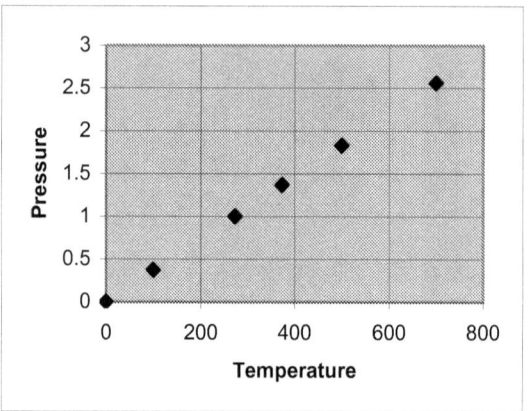

Figure 2.

At the freezing point of water, the pressure in the container is one unit, while it is slightly less than one and a half units at the boiling point. At higher or lower temperatures, the pressure is plotted against that temperature. As can be seen in the figure, there is a temperature at which the pressure would be zero. We can find this temperature is 273^0 C lower than the freezing point of water. We are justified in taking this temperature as the *absolute zero of temperature*. In fact, with any real gas, we cannot cool it sufficiently to reach this absolute zero where its pressure vanishes. All real gases will deviate from the behavior shown in Figure 2 as the temperature approaches -273^0 C. We can think of an *ideal gas* as a *gas that would have no pressure at absolute zero*. William Thomson, later named Lord Kelvin, proposed an absolute temperature scale in1848 based on this concept of absolute zero. This scale that uses the Celsius degree as its basic unit and a temperature zero at -273^0 C is now called the Kelvin scale and is the standard scientific temperature measure.

We have now quantified our ideas of relative hotness or coldness, but we have not yet made a connection between temperature and heat. In order to do this, we need to investigate the variation of temperature as heat enters or leaves a body. Consider taking a block of some material, for instance, aluminum. Attach a thermometer of some type to this block of aluminum and watch it as the aluminum block is held over a candle flame. We notice the temperature of the block rises uniformly with the time the block is held over the flame. Let us now take an aluminum block that is twice as massive as the first block and repeat our experiment. We notice the temperature rises at half the rate as in the first case. We assume the candle provides the same rate of heat flow in both instances so we conclude there is a direct proportionality between the heat flow and the product of the body's mass and temperature rise.

If we repeat the experiment with a different material, for instance a copper block, we notice exactly the same behavior as with the aluminum sample except the time needed for a certain temperature rise in identical blocks of aluminum and copper differs. We can look at the constant of proportionality that exists between the heat that flows into a metal block and the product of its mass and temperature change as a parameter that is unique to the particular metal. We call this proportionality constant the metal's *specific heat*. We generally use the letter, Q, to designate heat and, c, to designate specific heat. We now can write an equation relating these quantities:

$$Q = mc\Delta T \qquad \text{Equation 1.}$$

Historically, before we realized that heat and energy were the same, we devised units to quantify heat. These heat units, the calorie and the British Thermal Unit (BTU) are still in common use. We can define the specific heat of a particular material to establish the magnitude of a heat unit. The material we use is water. We define water as having a specific heat of one calorie per gram degree Celsius or one BTU per pound degree Fahrenheit. Therefore, the *calorie* is, by definition, *the amount of heat needed to raise the temperature of one gram of water one degree Celsius.* Likewise, *the BTU is the amount of heat needed to raise the temperature of one pound of water one degree Fahrenheit.*

Example 1. An outdoor grill burner is rated as producing 28,000 BTU per hour. How long would it take such a grill to raise the temperature of 10 pounds of tap water at an initial temperature of

55^0 F to the boiling point? Because water's normal boiling point is 212^0 F, the required temperature rise is 157^0 and the required heat is 1,570 BTU. At the stated rate, if all the heat generated by the grill went into heating the water it would take 3.36 minutes.

The relationship expressed in Equation 1 is valid only when the material remains in one state. When a change of state occurs, e.g. melting from a solid state to a liquid state, or vaporizing from a liquid state to a gaseous state, the heat that flows into the material goes into the internal energy of the material.

Consider taking a cube of ice at a temperature of -10^0 C and heating it with the same candle we used to heat the aluminum and copper blocks. As in the previous experiment, the ice block heats up until it reaches a temperature of 0^0 C. At that time, the temperature remains constant and the ice begins to melt. If we catch all the melted ice in a container, we find the liquid water remains at 0^0 C until the ice has completely melted. Once the entire ice block has melted, the liquid water's temperature begins to rise until all the water comes to a temperature of 100^0 C. Now, any additional heat goes into vaporizing the liquid water into a steam. If we close the system to catch all the steam, we find the temperature does not rise again until all the water is converted into steam. After this transition occurs, the temperature of the steam can again rise. The temperature as a function of time is shown in Figure 3.

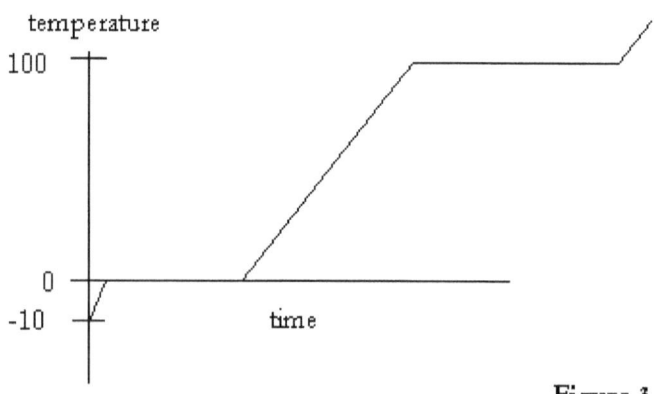

Figure 3.

An examination of Figure 3 leads to the conclusion that it takes much more heat to boil a given mass of water than it takes to melt that mass of ice. *The heat needed to change the state of matter is called latent heat.* The *latent heat of fusion* i.e. the heat needed to be removed from a mass of water at 0^0 C to convert it to ice or the amount of heat needed to convert a given mass of ice at 0^0 C to liquid water is approximately 80 calories per gram. The *latent heat of vaporization* is approximately 540 calories per gram. It takes an almost seven times greater amount of energy to convert a gram of liquid water into vapor than it takes to convert a gram of solid water into a liquid. There seems to be something fundamental in this large difference. We will consider this later.

Let's now consider the interrelationship of heat and mechanical work in more detail. Formally, this study is called *thermodynamics.* Thermodynamics is codified with three fundamental laws that were formulated during the nineteenth century. These laws relate the way heat and mechanical energy are changed from one to the other. A device that transforms heat into mechanical energy is called a *heat engine.* It was the desire to build heat engines during the nineteenth century that gave impetus to the development of thermodynamics.

Thermodynamics investigates the thermal interaction between a *system* and its *surroundings.* Specifically, the *First Law of Thermodynamics* investigates the relationship between the *internal energy* of a system, the *heat* that flows into or out of the system from the surroundings, and the *work* done by the system or on the system by the surroundings. The First Law states: *The change of internal energy of a system equals the heat that flows into the system minus the work the system does on its surroundings.* In fact, this statement applies to an idealized situation where the system undergoes reversible processes and dissipative forces are neglected. However, it is a useful starting point for the analysis of real thermal processes. In the form of an equation, the first law can be written:

$$\Delta U = Q - W \qquad \text{Equation 2.}$$

It is useful to consider a system consisting of an ideal gas contained in a cylinder fitted with a movable piston. In such a system, heat can flow into or out of the gas, and the piston can compress the gas, or expansion of the gas can cause the piston to perform work against the external surroundings. The process the system undergoes in moving from one set of conditions to another is important in the determination of the heat flow and work done, however, is irrelevant in the change of internal energy that occurs.

Formally, we say heat and work are *process dependent* variables while internal energy is a *process independent* variable.

The process independent character of internal energy means a system that undergoes a cyclic process will experience no change in internal energy. Heat engines are devices that take systems through multiple cycles where work is done against the surroundings. We define the *efficiency* of a heat engine as the *ratio of the work done by a device to the heat that flows into the device.* The *Second Law of Thermodynamics* deals with the efficiency of heat engines. In 1824, Sadi Carnot investigated the cyclic process that would lead to the most efficient possible engine. Carnot reasoned an engine operating between two heat reservoirs maintained at a higher and lower temperature would be most efficient if it followed a four-step cycle. That cycle consisted of an expansion at constant temperature where heat flowed into the system from the higher temperature reservoir, followed by a further expansion where no heat flowed, then a constant temperature compression followed by a compression with no heat flow.

Constant temperature, or *isothermal* processes are important in that the internal energy of the system does not change in the process. The internal energy of a system is directly proportional to its absolute temperature. Processes in which there is no heat flow are called *adiabatic* processes. The Carnot cycle consists of two isothermal legs and two adiabatic legs. Because all the heat flow is either at the temperature of the low or high temperature reservoir, the efficiency of a Carnot engine can be written as the ratio of the difference in temperatures to the temperature of the high temperature reservoir as shown in Equation 3.

$$\text{eff.} = \frac{T_H - T_C}{T_H} \qquad \text{Equation 3.}$$

Because the Carnot engine is the most efficient engine that can be made, we can use its efficiency equation to state the Second Law of Thermodynamics: *It is impossible to devise an engine that is 100% efficient unless the exhaust temperature is absolute zero.* We immediately state the Third Law of Thermodynamics: *It is impossible to attain an absolute zero of temperature.*

Example 2. What is the efficiency of a Carnot engine operating between water's boiling point and its freezing point? In order to calculate efficiency, we must use an absolute temperature scale. On the Kelvin temperature scale, water freezes at 273 K and boils at 373 K. Therrefore, the efficiency of a Carnot engine operating between these temperatures is 100 divided by 373, or about 27%.

The laws of thermodynamics are macroscopic laws in that they deal with gross physical properties of matter. While they were very important to the industrial development of the nineteenth century, they did not delve into the microscopic mechanisms that were the basis of macroscopic phenomena. Work done by James Maxwell and Ludwig Boltzman in the mid to late eighteen hundreds into the motion of the microscopic constituents of gases related the macroscopic thermodynamic laws to microscopic random motion in the *kinetic theory of gases.* We will discuss this theory at a later point.

Now, however, we can look at another phenomenon where heat is generated. In Chapter III we discussed connecting a conductor to a battery in order to have a path for the flow of electric current. By means of a chemical reaction, the battery caused the flow of electric charge through the wire. If we measure the temperature of the wire, we find it gets warmer the longer the current flows. This leads us to the conclusion that energy of some sort is being dissipated in the wire and is appearing in the form of heat. We need to investigate the source of the energy and the reason it appears as heat.

To investigate the source of this energy, we can make use of the current measuring instrument, the *galvanometer,* we mentioned in Chapter III. If we connect one wire to a battery terminal, the other end of it to one side of the galvanometer, then take another wire connected to the opposite side of the galvanometer and the second battery terminal, we provide a conducting path where charge can flow from the battery and return to it. Such a closed conducting path is called an *electric circuit.* The battery acts as the source of charge and provides the impetus for the movement of the charge through the circuit. Because charge is moves from the battery, it must have acquired some energy from the battery. This energy comes from the chemical reaction that occurs in the battery. Batteries that use different chemical reactions impart different amounts of energy to the charges they move. We find it useful to define *the energy per unit charge* that arises in the battery as the *electric potential* produced in the battery. The standard unit of potential is the joule per coulomb. This unit is named the *volt* in honor of Alessandro Volta who published the results of his investigations on chemical batteries in 1796.

In 1827, Georg Ohm presented the results of his investigations into the relationship between the current in a circuit and the voltage produced by the battery. Ohm connected various types of materials between the terminals of a battery. He reasoned the voltage produced by the battery must equal to the voltage lost or dropped across the material between the terminals. He observed a direct proportionality between the current in a material and the voltage drop across it. His results are illustrated in Figure 4.

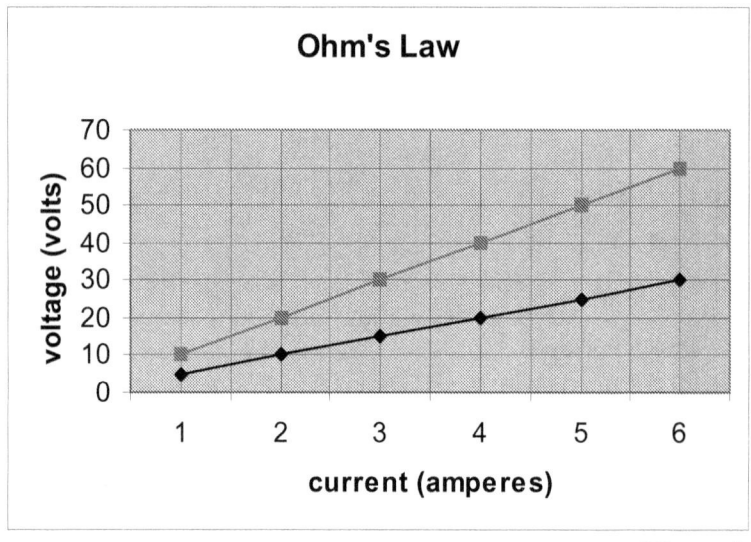

Figure 4.

The two lines shown in the figure represent the results of two different materials or of one material with different dimensions. The slope of the line, which represents the proportionality constant between the current and voltage, is the *resistance* of the material and is determined by the material type and its dimensions. The quantitative relationship between current and voltage is given in Equation 4.

$$V = IR \hspace{3cm} \text{Equation 4.}$$

The truth expressed in this equation is called *Ohm's Law*. Any material that always obeys this law is called a linear resistor.

While some materials, such as the filament of a light bulb, do not follow this linear relationship, we can always define the resistance of these non-linear materials as the ratio of the voltage drop across the material to the current through the material.

Example 3. What is the resistance of a material if a battery that produces 1.5 volts causes a current of 30 milliamps to flow through the material? The resistance is just the ratio of the applied voltage to the current that flows. In this case, 1.5 volts divided by 0.030 amps, or 50 ohms.

Because the voltage is energy per charge and current is a measure of charge per time, the product of current and voltage is the energy per unit time dissipated in a resistor. The *time rate of change of energy is called power*. We now have another way to demonstrate heat is just energy. We can take a container with a know mass of water and immerse a resistor in it. If we connect this resistor to a know voltage source and measure the current through the resistor for a given time, we quantitatively know the energy dissipated by the resistor. This energy appears in heating the water so, by measuring the temperature rise of the water, we again can determine the relationship between energy measured in joules and heat measured in calories. Numerically, the relationship is one calorie is equivalent to 4.186 joules. Once again, energy is conserved. The chemical energy in the battery was converted to electrical energy that was dissipated in the resistor and became heat energy. We can make the general statement of energy conservation that *energy is never created or destroyed, only its form changes*.

QUESTIONS

1. How was Joule able to demonstrate heat is just another form of energy?
2. What did Celsius do to overcome the difficulties of Fahrenheit's temperature scale?
3. How does the size of Fahrenheit's degree of temperature compare to Celsius's?
4. What is the inherent difficulty with both Fahrenheit and Celsius temperature scales?
5. How is the absolute zero of temperature determined?
6. How does an ideal gas differ from any real gas?
7. How does heat relate to temperature change in a material?
8. How is the calorie defined?
9. What is the specific heat of a material?
10. What do we mean by "latent heat"?
11. What can we conclude from the observation that the latent heat of fusion is less than the latent heat of vaporization?
12. Sometimes the First Law of Thermodynamics is stated, "You can't win." Why?
13. Sometimes the Second Law of Thermodynamics is stated, "You can't break even." Why?
14. Why can we write the efficiency of a Carnot engine in terms of temperature rather than heat?
15. What is the relationship between voltage and current in a linear resistor?
16. How is power defined?
17. How can we demonstrate the equivalence of electrical energy and heat?
18. What is the general statement of energy conservation?

CHAPTER VI

MOMENTUM

With the knowledge we gained in the previous chapters, we might now be tempted to think we can solve any problem that comes our way with just force and energy considerations. However, when we consider some problems, including the problems of collisions, we are not yet completely able to determine the outcomes. We need another tool to help us. Let's consider the head on collision of two very hard steel ball bearings as shown in Figure 1.

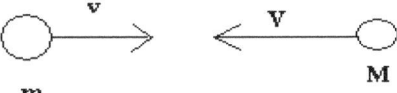

Figure 1.

The one ball of mass, m, and velocity, v, will collide with the second ball of mass, M, and velocity, V. After the collision, they recede from each other with velocities, v′ and V′ respectively, as shown in Figure 2.

Figure 2.

In general, v does not equal v′ and V does not equal V′. If this collision takes place where there is no difference in potential energies and no energy is lost in heat or deformation of the steel balls, the total kinetic energy of the system must be the same after the collision as it was before the collision. However, the kinetic energy of each body differs from its pre-collision value.

Now consider the situation where a steel ball collides with a stationary wooden block as shown in Figure 3.

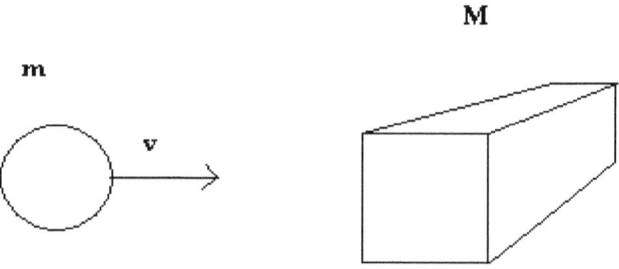

Figure 3.

After the collision, the ball can be imbedded in the block and both move off to the right with some velocity, V, which is much less than v, as shown in Figure 4.

M

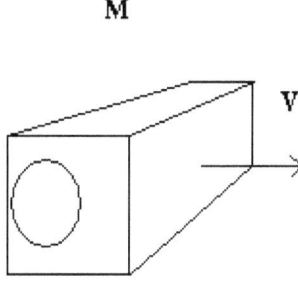

Figure 4.

In this case, the total kinetic energy   before collision is ½ mv^2 and after the collision is ½ $(m + M)V^2$. Because V is so much smaller than v, some kinetic energy is lost in the collision. But, no external forces acted during the collision so there must be some quantity that has not changed in the process. This quantity is the total *momentum* of the system.

Momentum is the *product of a particle's mass and its velocity*. As such, momentum is a vector quantity specified by a magnitude and a direction. We can define this in an equation.

$$\mathbf{p} = mv.$$ Equation 1.

In both collisions we observed, total momentum was conserved, however, total kinetic energy was conserved only in one of the collisions. We classify collisions where the total kinetic energy is conserved as *elastic collisions*, and those where there is a loss of kinetic energy as *inelastic collisions*. In the case where the two colliding bodies become one and move off together, the inelastic collision is called *perfectly inelastic*. In the case where there is absolutely no loss of kinetic energy, the collision is called *perfectly elastic*. In the macroscopic world, all collisions have some degree of inelasticity because some kinetic energy is lost. However, some materials, such as the material used to make "super balls", are very elastic and will return almost all the energy used to deform it during a collision. It is only when we enter the microscopic world that we find examples of essentially perfect elastic collisions.

Example 1. A 100gram projectile moving at 100 m/s collides with and embeds in a stationery 1 kg wood block. What is the magnitude of the velocity of the combined bodies after the collision? How much kinetic energy has been lost in the collision process? Because no external force acts during the collision, the momentum of the projectile before the collision must equal the momentum of the combined masses after the collision. Therefore, the combined masses must move at 9.1 m/s. Thus, the kinetic energy after the collision is 45.5 joules. Before the collision, the projectile had kinetic energy of 500 joules. About 90% of the initial kinetic energy was lost.

Consider, for example, the collisions of the molecules of a gas confined to a cylinder fitted with a movable piston as shown in Figure 5.

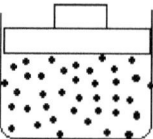

Figure 5.

Each of the molecules collides with the sides of the cylinder or the piston at the top of the cylinder. If we think of the gas as being composed of point particles, the only collisions that are possible are those with the container. In the last chapter we discussed the internal energy of a system. If that system is a gas composed of these non-interacting molecules, the internal energy of the gas is just the sum of the kinetic energies of the moving molecules. If the gas is compressed as the piston is moved down by the application of an external force, work is done on the system. If the gas expands and forces the piston up, it does work against the surroundings. Therefore, there must be some connection between the movement and collisions of these gas molecules and the macroscopic thermodynamic variables we discussed previously. This was Boltzman's conclusion that led to the formulation of the *kinetic theory of gases.*

Consider the collision of one molecule with the piston. The molecule hits the piston and rebounds. In the collision process, the molecule's momentum has changed because its direction of motion differed before and after the collision. This momentum change occurred because some force acted on the molecule. That was the force of the piston on the molecule during the collision.

This is the same type of force that acts on a tennis ball when it collides with a racket, or a baseball that collides with a bat. In these instances, the force acts while the ball is in contact with the racket or the bat. The product of the force and the time it acts is called *impulse.* This impulse caused a change of momentum. We can now restate Newton's Second Law of motion by stating *a force on an object produces a time rate of change of its momentum.* This statement is entirely consistent with the statement we made that force acting on a mass causes the mass to accelerate, as long as the mass remains constant.

Making use of Newton's third law, we are forced to conclude the force of the piston on the molecule during the collision must be equal and opposite to the force of the molecule on the piston. This being true, we can determine the average force on the piston caused by every molecule's collision with it. If we know the time it takes for the molecule to leave the piston face, travel to the cylinder's bottom, bounce off the bottom, and again travel to the piston to collide again, we can measure the force of this molecule on the piston. However, that is impossible because we cannot identify one molecule and follow its motion. If we have many molecules, we can consider the average molecular motion and use statistics to investigate the problem. The problem now becomes one of determining, the average velocities of the molecules.

Maxwell and Boltzman knew that all molecules do not move with the same velocity, but there is a statistical distribution of velocities. This distribution is mathematically described by an equation that relates the number of molecules at a given velocity to that velocity. In this distribution, we can discuss the mean or average speed of the particles, the median speed of the particles or the rms (root mean squared) speed of the particles. Because the average kinetic energy of the molecules is related to the average velocity squared, this latter value gives us most information.

Consider again the collisions of the molecules with the piston. Knowing something about the average velocities of the molecules and, therefore, the time between collisions, we need simply to know the number of molecules in a gas contained in the cylinder to determine the force on the piston. The ratio of the force on the piston to its area is, by definition, the gas *pressure*. But, Amedeo Avogrado in 1811 determined the number of molecules in a given molecular weight of material. Therefore, if we know the mass of the gas we have in the cylinder and its molecular weight, we know how many gas molecules are contained. We now have a linkage between the microscopic motion of molecules and a macroscopic variable, the gas pressure.

The Maxwell-Boltzman distribution function, shown in Figure 6, is dependent on temperature. The higher the temperature of the system of molecules, the greater is the number of molecules that move with a higher velocity. The curve shown in Figure 6 is that of a gas at relatively low temperature. As the temperature rises, the curve flattens out and the peak shifts to higher speed so that the area under the curve that represents the total number of molecules in the container remains constant.

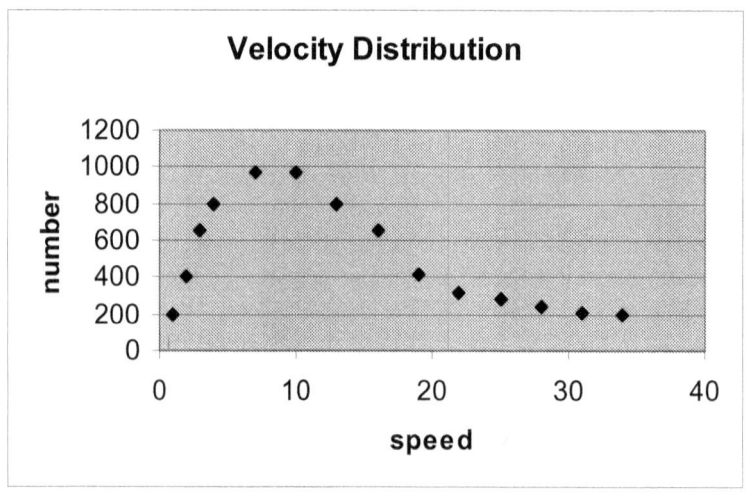

Figure 6.

FUNDAMENTAL CONCEPTS OF PHYSICS

Here is another linkage between a macroscopic parameter, the temperature, and the microscopic motion of the molecules. The entire kinetic theory of gases brings together the macroscopic with the microscopic.

Let's return to the macroscopic world for some considerations of the concept of momentum and its conservation in our everyday observations. Consider a fireworks display where explosive shells are launched into the air and exploded. If the shell does not explode, it follows a parabolic trajectory as we described in Chapter II. The explosion that occurs is due to a chemical reaction that takes place within the shell, not the action of any external force. Therefore, since no external force acts, the total momentum of the shell must be the same after the explosion as it is before the explosion. In fact, this is the case. If we measured the momentum of all the shell fragments after the explosion and added their individual momentum vectors, we would find the total momentum of all the pieces just adds up to the momentum initially possessed by the shell. Momentum is conserved because no external force acts.

Likewise if we look at the shooting of a shotgun, we find the total momentum of the shotgun and the shot expelled from it after the trigger is pulled is zero. Before the gun is fired it is at rest. After the trigger was pulled, shot, carrying momentum, leaves the muzzle causing the gun to recoil against the shooter's shoulder. This is perfectly reasonable because no external force acts and any momentum imparted to the shot must be balanced by opposite momentum imparted to the gun. Momentum is conserved.

Obviously, the concepts of energy and momentum are very powerful tools in helping us gain understanding of the workings of the physical world on a microscopic and a macroscopic level. However, until now the problems we investigated were a problems involving linear motion. We need to consider situations where objects move in curved paths and not straight lines.

Consider a situation where an external force acts on a body to alter the body's momentum but not change its speed. Such a situation occurs when we consider the problem of *uniform circular motion*. In this type of motion an object moves in a circular path with constant speed. This motion is, in fact, simpler than the motion of the planets that we investigated in the first chapter. The planets traveled in orbits where the direction and the speed of the planet continuously changed. The problem at hand considers only a change in direction.

Michael J. Cardamone

Certainly, there must be a force that causes the direction of the velocity vector to change. This force is called a *centripetal*, or center-seeking force. Any force that only changes a velocity's direction and not it magnitude is a centripetal force. By Newton's Second Law, there must be an acceleration associated with the centripetal force. This is the centripetal acceleration. The question now before us concerns the quantification of centripetal forces and accelerations. How can we measure these quantities?

One method is to attach a weight to a string and swing the string in a way that the weight moves in a horizontal circle. The force of the string on the weight is the centripetal force in this case causing the centripetal acceleration. If we spin the weight at a relatively low speed and release it, we can determine the magnitude of its velocity when released by measuring its trajectory. We can repeat this observation with differing speeds of rotation. Now, let's put a spring scale between the weight and the string to measure the force as well as the speed. When we perform this experiment we discover that the centripetal force is related to the square of the weight's speed. Repeating the experiment with longer or shorter pieces of string, we observe the centripetal force varies as the reciprocal of the string's length. Putting all our information together, we conclude the centripetal force is proportional to the ratio of the speed squared to the radius of the circular path described by the weight. We identify this ratio as the *centripetal* acceleration. As was the case with linear motion, the constant of proportionality between the force and the acceleration is the mass of the moving object. Thus we are able to write the centripetal force as

$$F_c = mv^2/r \qquad \text{Equation 2.}$$

Returning for a moment to problems such as those of planetary motion where the planet's speed and direction continuously change under the action of the gravity, we need to specify not only the planet's *centripetal*, acceleration, but also the acceleration associated with it change in speed. This acceleration is called its *tangential* acceleration because the force that causes this acceleration acts in a direction parallel, or tangent, to the direction of the particle's velocity. In general, whenever a force acts on a body we can consider the force to have a centripetal and a tangential component. The centripetal component only changes the particle's direction, and the tangential component only changes the particle's speed.

It is interesting to note that a centripetal force affects a particle's momentum but not its kinetic energy. The reason is, of course, momentum is a vector quantity while energy is scalar.

Now that we have introduced the concept of circular motion, we need to investigate the kinematics of this motion as we did with linear motion. In the linear case, we described the linear displacement of a particle from one point to another. In the case of circular motion, we can describe the *angular* displacement of a particle following a circular path as the angle swept out by the line joining the particle to the center of curvature, the radius of the circular path. Generally, this angular displacement is identified by the Greek symbol θ. Following standard geometric definitions, this angle is defined as the ratio of the linear distance the particle has moved on its curved path, S, to the circular path's radius.

$$\theta = S / r \qquad \text{Equation 3.}$$

If we take the time rate of change of the angular displacement to get the angular velocity designated by ω, we obtain the relation between this and the linear velocity

$$\omega = v / r \qquad \text{Equation 4.}$$

Further differentiation with respect to time yields the relation between angular and linear acceleration, α.

These angular variables follow kinematics in the same way the linear variables of displacement, velocity and acceleration do. Likewise, the dynamical variables of force, momentum, and energy have angular counterparts. Notice all the linear and angular variables are related through the radius of the curved path followed by the particle. If we consider a body that is extended in space, every particle in the extended body will be following a different radius when the body is set into circular motion. Therefore, when we consider the action of forces on such bodies, we need to consider the distance from the center of rotation to the particular massive particle being affected. To accurately measure the action of forces on the motion of extended bodies, we consider the distribution of the mass by defining the *moment of inertia*, I, of the body. In the angular motion of an extended body, this quantity takes the place of the mass in our considerations. The force acting through a distance on an extended body is called *torque*, τ, and is mathematically defined as the vector product of the distance through which a force acts and the force. In the same way that forces cause masses to accelerate, torques cause extended masses to undergo angular acceleration. For rotational motion, Newton's Second Law of Motion is written

$$\tau = I\alpha \qquad\qquad \text{Equation 5.}$$

A rotating body possesses angular momentum the same as a linearly moving body has linear momentum. The angular momentum of a rotating body, L, is simply the product of its moment of inertia and its angular velocity

$$L = I\,\omega. \qquad\qquad \text{Equation 6.}$$

Again as in the case of linear motion, angular momentum is conserved if no external torque acts. This fact is illustrated in observation of the spin of a figure skater. When the skater's arms are extended, she spins with some angular velocity. When she pulls her arms into her body, her moment of inertia becomes smaller. But, no external torque has acted on her, so her angular velocity increases and she spins much faster.

QUESTIONS

1. What quantities are conserved in an elastic collision if no external force acts?
2. What quantities are conserved in an inelastic collision if no external force acts?
3. Can an object possess energy but not momentum? Why?
4. Can an object possess momentum but not energy? Why?
5. How is the pressure of a gas, a macroscopic variable, related to the velocity of the gas molecules, microscopic variables?
6. Where is the internal energy of a gas contained according to the kinetic theory of gases?
7. What was Avogadro's contribution to our understanding of gases?
8. Why must the area under the curve in the Maxwell-Boltzman velocity distribution graph remain the same for a given amount of gas regardless of temperature?
9. What happens to the peak of the Maxwell-Boltzman velocity distribution graph as the temperature of the gas rises?
10. How could you use the concept of momentum conservation to explain the flight of a rocket?
11. What is meant by a centripetal force?
12. How do centripetal and tangential accelerations differ?
13. Why does a centripetal force not affect a particle's energy?
14. How are the linear variables of displacement and velocity related to the angular variables of angular displacement and angular velocity?
15. What is the angular analog to the linear concept of mass?
16. State Newton's Laws of Motion for angular motion.
17. What is angular momentum?

CHAPTER VII

WAVES

Consider what happens when you drop a stone into a pool of water. The stone has some kinetic energy when it strikes the water surface. Some of this energy causes ripples on the water's surface to spread from the point the stone entered the water. If we have a cork float in the water at some distance from the stone's entry point the ripples will eventually cause the float to oscillate. Obviously, the float has acquired energy that was originally possessed by the stone. The energy was transmitted from the place the stone entered the pool to the place the cork was floating although there was no motion of any physical body from one point to the other. The mechanism for the transmission of the energy was the water *wave* produced when the stone entered the pool. One thing characteristic of all waves is their ability to *transport energy from one point of space to another without the physical motion of any object from the first point to the second.*

In the present situation, energy from the stone entered the water and traveled in the water to the position of the float. In this instance, the water was the *medium* through which the energy propagated. No individual particle of the medium moved from the position of the stone to the float, but the energy traveled from one point to the other. If we measure the distance between the stone and the float, and the time it takes for the energy to get from one point to the other, we can measure the wave's *velocity.*

Let's repeat the same experiment of dropping a stone into a pool, but this time we will fill the pool with molasses instead of water. When we again measure the wave's velocity, we discover it is different than it was in water. We can use other liquids and repeat the experiment several times to reach the conclusion that the speed of a wave is dependent on the medium through which it moves.

Let's now investigate a string attached to a tuning fork and passed over a pulley as shown in Figure 1.

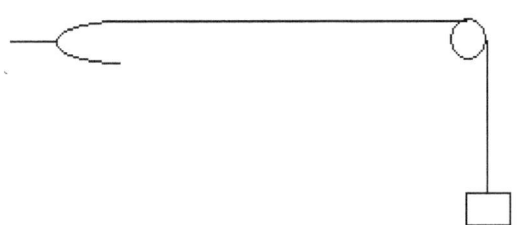

Figure 1.

The string is held tight by the hanging mass attached to it. When struck, the tuning fork vibrates with a fixed *frequency, the number of vibrations it makes per unit time.* This idea of frequency is the same one we introduced when discussing simple harmonic motion. In fact, we can say the end of the string attached to the tuning fork is undergoing simple harmonic motion in a direction perpendicular to the length of the string.

Each particle of the string affects the particle next to it in a way that every individual element of the string is vibrating with the frequency imposed by the tuning fork in a direction perpendicular to the direction in which energy is propagated along the string. The wave so produced on the string is called a *transverse wave.* Any wave where the particles of the medium vibrate in a direction perpendicular to the direction of energy transport is a transverse wave.

If we hold the tuning fork fixed and pluck the string to view the propagation of one transverse wave pulse down the string, we can investigate how the weight hanging on the string holding it under tension effects the wave pulse's speed. When we add more weight to the end of the string, we find the speed of a wave pulse increases. If we quantify this, we find the wave speed is proportional to the square root of the weight, or string tension. Further investigations with other strings, leads us to the conclusion that the square of the wave speed is directly proportional to the ratio of the tension on the string to its linear density, i.e. its mass per length. We can write this result in an equation as:

$$v^2 = T/(m/l) \qquad\qquad \text{Equation 1.}$$

Let's now attach our string to a fixed wall and pluck it, as shown in Figure 2.

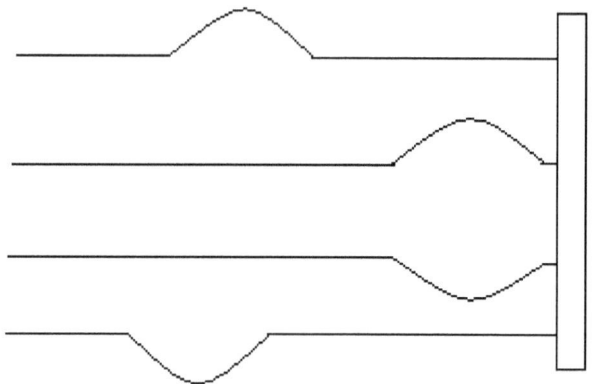

Figure 2.

We notice the wave pulse travels to the right until it strikes the wall, is reflected at the wall, and travels back to the left. We can put two successive wave pulses on the string and observe them as the pulse traveling down the string and the pulse reflected at the wall traveling back to the left pass along the same length of string. Figure 3 shows the two pulses pass through each other in such a way that after they have passed the length where they are coincident their properties are the same as they were before they met.

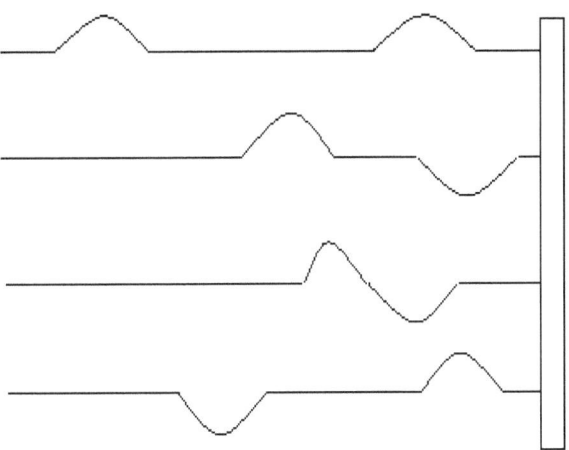

Figure 3.

When they are at the same position on the string, the transverse displacement of each of the string particles is the sum of the displacement the particle would have experienced from each wave pulse individually. This is an example of the principle of *superposition of waves.* Simply put, this principle states *the total displacement of a medium particle is the sum of the displacements the particle would experience due to the presence of a number of waves.* This principle is of great importance in that it explains how waves are able to *interfere* with each other. If two waves come together so that the displacements due to each are in the same direction, the total displacement will be the sum of the individual displacements. If the individual displacements are in opposite directions, the total displacement will be the difference of the individual displacements.

We can now continuously drive our tuning fork to produce continuous waves on the string, we can adjust the weight to produce a pattern where the string appears not to move at all at some points and vibrates with a maximum amplitude midway between these points. Such a pattern is called a standing wave pattern and is produced by the superposition of waves traveling down the string and back up. The positions where there is no motion of the string are called *nodes* of the standing wave pattern and the position of maximum amplitude are called *antinodes*.

A detailed analysis of the standing wave pattern leads us to conclude waves on the string extend in space for a certain distance then repeat their pattern. This spatial extent of a wave before its displacement pattern repeats is called its *wavelength,* and is usually designated by the Greek letter lambda, λ.

In this case, as in every instance with a wave, the product of a wave's wavelength and frequency is its velocity. Again, in the form of an equation, this is written:

$$v = \lambda f \qquad\qquad \text{Equation 2.}$$

If we replace our string with a spring that is attached to the wall, we can hit the spring to compress it along its length. This compressive pulse we have created travels the length of the spring, hits the wall, and travels back, in much the same way as the wave pulse traveled on the string. The difference between the two situations is that in the case of the string particles of the string executed SHO in a direction perpendicular to the direction of energy propagation, while in the instance of the spring, the spring coils vibrated *in* the direction of energy propagation. If we drive the end of the spring so as to continuously place compression pulses on the spring, we will create a *longitudinal wave* in the spring. *A longitudinal wave is a wave where particles of the medium undergo oscillation in the direction of energy propagation.*

Longitudinal waves follow the same rules as transverse waves. The frequency of a longitudinal wave is the number of compressions per unit time the wave performs. The wavelength is the distance between repetitions of the compression. A medium is needed for the propagation of the wave. The velocity of the wave is determined by the properties of the medium. The product of the longitudinal wave's frequency and its wavelength equals its speed.

Regardless of the transverse or longitudinal character of waves, the principle of superposition of waves holds. This principle is useful in explaining the effects of several waves coming together at some specific point in space and time. One characteristic of transverse waves that is not true of longitudinal waves is that transverse waves may be *plane polarized*, i.e. *all particles in the medium vibrate in the same plane that is perpendicular to the direction of propagation.*

When waves come together at the same point in space and time their effects can add or cancel depending on the particular position of the waves peaks and valleys. *If waves come together so that their crests and dips are coincident, the waves are said to interfere constructively. If they come together so that the crests of one wave fit into the valleys of another, they interfere destructively.* Destructive interference is shown in Figure 4.

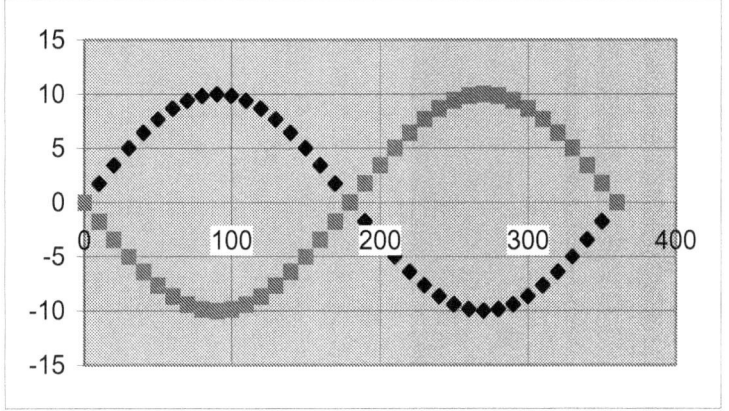

Figure 4.

Notice that the sum of the disturbance at any point is identically equal to zero.

The constructive interference of two waves with differing amplitude is shown in Figure 5.

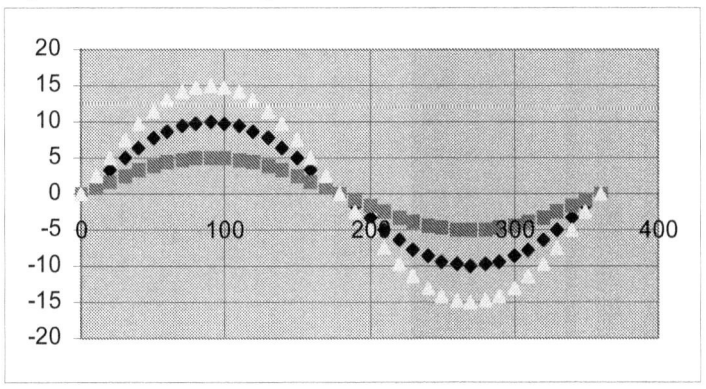

Figure 5.

In this situation, a wave with amplitude 10 units and one with amplitude 5 units interfere constructively to form one wave of amplitude 15 units.

The fact that we observe interference effects with sound leads us to believe that sound is a wave phenomenon. When a musical instrument is being tuned, a particular note on the instrument is struck while at the same time a note of known pitch is produced by some standard sound source. If the instrument is nearly in tune, the sound from the instrument and the sound from the standard source will add together to form one sound that varies in loudness. The instrument being tuned is adjusted until the loudness variation ceases. The loudness variation is called *beats,* and occurs with a *beat frequency,* which is the difference in the frequency of the sound produced by the instrument and the sound produced by the standard. This beat phenomenon is just an example of constructive and destructive interference in sound. Only waves can interfere, thus we must conclude that sound is a wave.

Sound waves are longitudinal waves in a material. Our ear is sensitive to periodic variations of air pressure in the frequency range of 20 Hz to 20,000 Hz. *The Hertz, Hz, is the frequency unit equal to one vibration per second.* We can demonstrate sound needs a medium for its transmission by placing an electric doorbell inside a jar that can have the air evacuated from it. When we ring the bell while the jar is still filled with air, we hear the ringing bell and see the striking of the clapper. When we evacuate the jar, we still see the action of the clapper, but no longer hear the sound of the bell. The air in the jar was the medium through which the sound wave propagated.

Our ear is the instrument that first introduces us to sound. As such, our first knowledge of sound is based on a physiological awareness. We need to relate the physical parameters of sound waves to the physiological observations we have made. The first thing we notice about sound is its relative loudness or softness. The physical variable associated with a sound's loudness is the *intensity* of the sound wave. *Intensity of a wave is defined as the energy per unit time the wave transports through a unit area.*

In addition to a sound's relative loudness or softness is its *pitch*, a measure of how high or low the sound is heard. Pitch is most closely associated with the sound's *frequency*. High pitch sound consists of higher frequencies than does low pitch sound. The musical note "A" is a sound wave with a frequency of 440 Hz while middle "C" is a sound wave of frequency 261.6 Hz.

FUNDAMENTAL CONCEPTS OF PHYSICS

The sound produced by a musical instrument consists not only of the fundamental frequency of the note produced, but also by *overtones,* which are waves with frequencies that are multiples of the fundamental frequency. The number and intensity of the overtone frequencies gives the sound from a particular instrument its unique sound. An "A" played on a flute does not sound identical to the same note played on a piano. The particular set of overtones in a sound gives the sound its *timbre.*

Until now, we have concentrated on sound sources and observers who are stationary. What would occur if they moved? How would the observed sound change? The Austrian scientist Christian Doppler explained this phenomenon in 1842. He noted that each successive pressure maximum a stationary observer would experience from a sound source approaching him would arrive at his location in a shorter time than the sound approaching him from a stationary source. Likewise, the pressure maxima would be observed with longer time intervals if the source were receding from the observer. Therefore, the observer would hear a higher frequency when a source approached and a lower frequency when the source receded. In the same fashion, if the source remained stationary and the observer moved, the sound observed would be of a higher frequency when the observer approached the source and lower when he departed form it. This *Doppler Effect* is a characteristic of all waves and is important in applications as varied as measurement of speeds of galaxies or pitches thrown by a baseball player.

QUESTIONS

1. What is a wave?
2. How do the particles in a medium move when a wave is propagated through it?
3. What is the difference between a transverse and a longitudinal wave?
4. What is meant by a wave's frequency?
5. What is meant by a wave's wavelength?
6. What determines the speed of a wave on a string?
7. What is the principle of superposition of waves?
8. What is necessary for waves to constructively interfere?
9. In a standing wave, what are nodes and antinodes?
10. What do we mean by plane polarization of a transverse wave?
11. What is the nature of sound waves?
12. How is a wave's speed related to its frequency and wavelength?
13. What is meant by beats?
14. How is the intensity of a sound wave related to the sound we hear?
15. Why do two different type musical instruments playing the same note not sound identical?
16. What is meant by the Doppler Effect?
17. How does the sound you hear from a train's horn change as it first approaches you and then recedes from you?

CHAPTER VIII

FIELDS

In our first effort to quantify mechanics, we discovered that force is one of the most fundamental concepts for our understanding of the interaction of one body with another. When we introduced the concept of force, we avoided any explicit statement of the mechanism with which one object at a distance from another object could affect its motion. This is, perhaps, more of a philosophical than a physics difficulty. However, it leads to a great deal of discussion that could be categorized into two general schools of thought, those who believed in "action at a distance" and those who believed in some medium through which force acted.

The "action at a distance" school of thought proposed that there was no necessity for a medium to transmit force and that the mere presence of one body could affect the motion of another body separated from it. Those who rejected this concept, particularly Michael Faraday, developed an alternative explanation during the second and third decades of the eighteenth century. Their explanation postulated the existence of a *force field*. Specifically, they stated that, *a force field is a modification of the space surrounding a body such that another body brought into the vicinity would experience a force.*

While this statement may, at first sight, seem intimidating, in reality the concept is not difficult. Consider the situation we discussed in Chapter III when we investigated the force between two charged bodies. This situation is illustrated in the following figure.

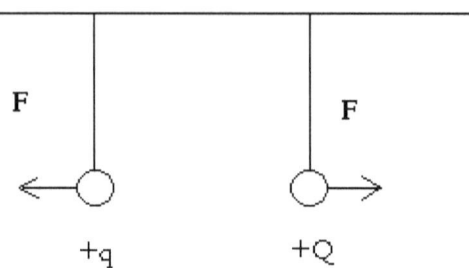

Figure 1.

We have already observed the two bodies repel each other with a force proportional to the product of the charges and inversely proportional to the square of the distance between them. In Figure 1, we can say there is a force on the body with charge +q due to the presence of the body with charge +Q. If we remove the +q charged body, have we done anything to the ability of the charge +Q to interact with another charged body?

We can investigate this by bringing another charged body to the position formerly occupied by the removed body. Consider bringing in an object with charge −2q as illustrated in Figure 2.

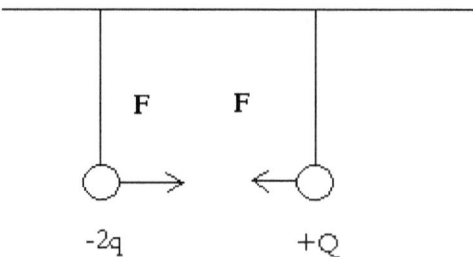

Figure 2.

In this situation, there is an attractive force between the two bodies that is twice a strong as was the repulsive force in the first case. From these experiments, we can conclude that the mere presence of the object with charge +Q modified the space at the site of the of the other body such that any charged body brought into that space would experience a force of attraction if it were negatively charged or repulsion if it were positively charged. These results are completely consistent with the definition of a force field given above. Because the objects we are dealing with interact through the electrostatic force, we can say that the presence of a charged body sets up an electrostatic force field around it such that any other charged body will experience an electrostatic force.

We know the form of the Universal Gravitation Law is the same as that of Coulomb's electrostatic force law, so it is reasonable for us to consider the situation we have just considered with the charged bodies replaced by massive bodies. When we investigate this situation, we again come to the conclusion that a massive body modifies the space surrounding it in such a way that another massive body will experience a gravitational force. The differences between the gravitational case and the electrostatic case is simply in the magnitude of the forces involved and the fact that electrostatic forces can be attractive or repulsive while gravitational forces are only attractive. We thus conclude that every massive body has a gravitational field associated with it.

To quantify our ideas about force fields, again consider the situation described in Figure 1. If we let the charge +q become extremely small, there would still be a modification of space at its location due to the charge +Q such that any charge brought to that point would experience a force that would be proportional to its charge magnitude. We can use this fact to quantitatively define the electrostatic field at this point in space as equal to the ratio of the force a test charge would experience at that point to the magnitude of that test charge. In terms of an equation, this can be written:

$$\mathbf{E} = \mathbf{F}/q \qquad \qquad \text{Equation 1.}$$

Because force is a vector quantity, the electrostatic field must also have a vector characteristic. We define the direction of this field, usually just called the electric field, as being in the direction a positive charge would be accelerated when located at that point in space. In the SI system of units, the units of the electric field are newtons per coulomb, or equivalently, volts per meter.

Example 1. What is the electric field at a point of space if a +5 coulomb charge experiences a repulsive force of 30 newtons? Because the positive test charge experiences a repulsive force, the electric field at this point is positive with a magnitude of 6 N/C.

In the gravitational case, we can use similar arguments to define the gravitational field at a point in space as the ratio of the gravitational force on a mass at some point in space to the magnitude of the mass:

$$G = F/m \qquad\qquad \text{Equation 2.}$$

The gravitational field has the units of newtons per kilogram.

With this quantification of the concept of a field, whenever we discuss the interactions that occur when a charged body moves due to the presence of other charged bodies or when a massive body moves due to the presence of other massive bodies we can discuss the motions in terms of the interaction of a material body with a field. This allows a much more fundamental understanding of physical processes that will be important in our future considerations.

With our new knowledge of fields, let's reinvestigate some of the magnetic phenomena we discovered in a previous chapter. We know when magnets are brought near each other they interact. Therefore, in much the same way as we defined the electrostatic field for a charge distribution we can define a magnetic field which exists for a permanent magnet. If we repeat Oersted's experiment discussed in Chapter III, we can take a bar magnet and investigate the field it produces by means of a small magnetic compass brought into the vicinity of the magnet. Thus, we can determine the field produced by the magnet. This situation is illustrated in Figure 3.

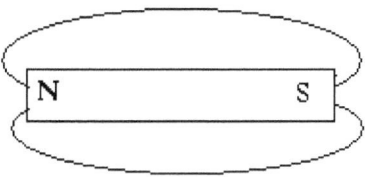

Figure 3.

The magnetic field traces out closed arcs about the magnet. This type of field pattern is similar to the electric field pattern that exists between two opposite electric charges. Such a combination of charges of equal magnitude and opposites sign separated by a specific distance is called an electric *dipole*. The field pattern of the bar magnet suggests it may be thought of as a *magnetic dipole*. No matter how small we make our bar magnet by breaking it into tiny pieces, each piece will reproduce the same field pattern leading us to conclude that nature does not permit the existence of *magnetic monopoles* or isolated *magnetic charges*, but only allows the existence of magnetic dipoles, consistent with the results we found in Chapter III when we broke a bar magnet into smaller and smaller pieces.

Knowing that a magnetic compass brought near a current carrying conductor experiences a magnetic force, it would be logical to deduce that the electric current is the source of the magnetic field. Or, since current is simply a measure of the motion of charge, that moving charges act to produce magnetic fields. Biot and Savart quantified the magnetic field produced by a moving charge in 1820.

Knowing now that electric currents are sources of magnetic field, and that moving charges in a region of magnetic field experience magnetic forces, we are ready to investigate the force between two current carrying conductors, illustrated in Figure 4.

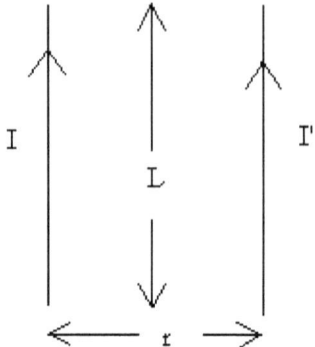

Figure 4.

Consider two current carrying conductors of length, L, separated by a distance, r. In the situation shown, one conductor carries current I and the other carries current I'. We can think of this situation as the conductor carrying current I sets up a magnetic field surrounding it so that the conductor carrying current I' experiences a force. Or, we can consider the conductor carrying I' sets up the field that causes the force on conductor with current I. Obviously, these two interpretations are consistent. Therefore we can discuss the force between current carrying conductors. Andre Ampere performed experiments investigating this phenomenon and published his results in 1822.

By measuring the force between the conductors, Ampere concluded the force per unit length of conductor was proportional to the product of the currents carried and inversely proportional to the distance between the conductors. If the currents were in the same direction, the force was attractive, if in opposite directions, the force was repulsive. These findings allowed us to define a unit of current in terms of the force per unit length between conductors. In the SI system of units we define the *ampere* as *the current in two conductors separated by a distance of one meter that will cause a force per unit length between the conductors of strength 2 x 10⁻⁷ newtons per meter.*

With the unit of current now quantified, we can return to the work of Biot and Savart to quantify our understanding of the magnetic field. In the SI system of units, the magnetic field is called the *tesla* in honor of Nikola Tesla. One tesla is equal to one newton per ampere meter.

Let's now consider the effect a magnetic field has on a single moving charged particle. Consider the magnetic field in some region of space. If a charged particle is inserted into this region, it will experience a force that is perpendicular to its velocity. This force is a *centripetal* force that will cause the particle to move in a circular path. This fact will be useful when we discuss the characteristics of atomic and sub-atomic particles.

The fields we discussed so far are all static in nature. They do not vary with time. As such, our knowledge is incomplete. It is also incomplete in that there is a certain lack of symmetry to it. We know that a current, or flow of electric charge, can create a magnetic field, but we have yet to see a changing magnetic field create a flow of charge. This observed lack of symmetry lead Michael Faraday to investigate the possibility of a moving magnetic field inducing a current in a conductor. He coiled some wire around a bar magnet and connected the wire to a galvanometer. While the magnet was at rest, the galvanometer recorded no current in the wire. However, when Faraday moved the magnet, a current was indicated. When he pushed the magnet farther into the coil, the current flowed in one direction. When he pulled the magnet out of the coiled wire, the current's direction was reversed. Faraday when he published his results in 1831 concluded that a *changing*, or *time varying*, magnetic field induced an electric current. To be a bit more precise, we can state *Faraday's Law* as*: A time varying magnetic field through a coil will induce an emf (electromotive force) which will produce a current in the coil.*

Approximately the same time Faraday was researching in England, Joseph Henry, the American physicist, was investigating induced currents. In fact, Henry's experiments worked much better than Faraday's. Henry took a piece soft iron shaped like a horseshoe and wrapped a wire around it to produce an electro-magnet, as shown in Figure 5. When current was passed through the wire, a magnetic field that followed the iron horseshoe was produced. Henry then connected another piece of iron across the opening of the horseshoe, wound wire across this piece of iron, and connected the wire to a galvanometer. When the wire around the primary horseshoe was connected to a battery so that a current could flow, Henry observed a momentary deflection of the galvanometer indicating the presence of a current in the secondary coil. The galvanometer returned to its equilibrium position until the circuit in the primary coil was broken stopping the flow of current. When this occurred, Henry noted another momentary galvanometer deflection.

Henry's observations and Faraday's results are consistent. They both indicate *a static magnetic field produces no current, only a varying magnetic field will induce a current.*

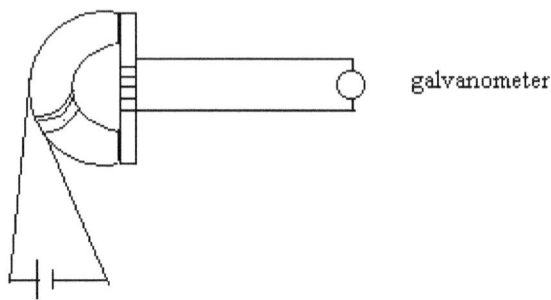

galvanometer

Figure 5.

We can make practical use of this phenomenon to produce a device that alters currents and voltages that vary in time. Currents that vary such that their average value over time is zero are called *alternating currents* or, *ac,* as opposed to currents that have an average value other than zero that are called *direct currents,* or *dc.* Direct currents are produced by the chemical reactions in batteries or by direct current *generators,* while *alternators* produce alternating currents.

If we now take a piece of iron as shown in Figure 6, wrap several turns of wire around one side, and several turns of wire on the other side, we can investigate the relationship of the current we take from an ac source and the current induced in the secondary coil.

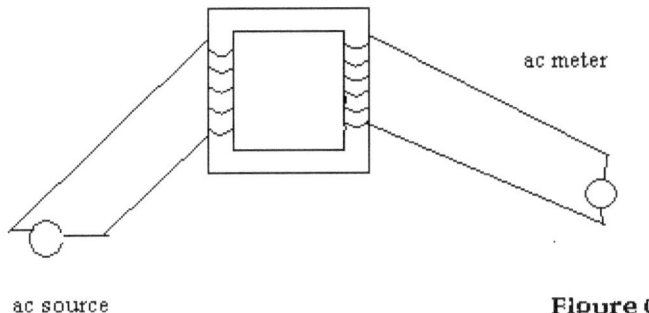

ac meter

ac source   **Figure 6.**

The time varying current in the first coil creates a time varying magnetic field that is contained in the iron core and goes through the second coil. This time varying field induces an ac current in the second coil. If we measure the ratio of the relative effective values of the primary current and the secondary current, we find it equals the inverse ratio of the number of turns of wire in the primary coil to the secondary coil. Because the power delivered to the primary coil is transmitted to the secondary coil, the product of voltage and current in the primary coil must equal the product of the current and voltage in the secondary coil. Therefore, the ratio of the voltage in the primary coil to that in the secondary coil equals the ratio of the number of turns in the primary coil to the number in the secondary coil. This device that *transforms* the value of the voltage is called a *transformer*. If a transformer has more turns in its secondary coil than in its primary coil, the secondary voltage is greater than primary voltage. This is a *step-up transformer*. If the opposite is the case, the voltage at the secondary is less than that at the primary and the transformer is a *step-down transformer*.

Example 2. A transformer is constructed with a primary coil consisting of 100 turns of wire and a secondary coil consisting of 12,000 turns of wire. If the primary coil is connected to a normal house outlet where the voltage is 110 volts, what voltage will be observed on the secondary coil? This step-up transformer will boost the voltage to 13,200 volts.

Let's return out attention to Henry's experiment shown in Figure 5, and examine the direction of the induced current detected by the galvanometer. This induced current, being a real flow of charge, produces a magnetic field. However, it was a changing magnetic field that was responsible for the induced current in the first place. If the magnetic field whose source is the induced current were in the same direction as the primary field, the induced field would itself produce more current that would produce more field which would induce more current and so on to infinity. This would be getting something for nothing. While this might solve all our energy problems, nature never acts in such a manner. Thus, we are forced to conclude the induced current must be in such a direction so as to set up a magnetic field that opposes rather than reinforces the primary field responsible for its existence. The Russian scientist, Heinrich Lenz, arrived at this conclusion in 1834. In his honor, we name *Lenz's Law: The direction of any effect produced by a magnetic induction process is in a direction such that it opposes the cause of the effect.* While Lenz's Law sounds very formal, in reality it is another statement of simple truth of energy conservation.

We are able to consider now the interrelationship between changing magnetic fields and changing electric fields. We know that a moving charge is affected by a magnetic field. Consider a moving charge affected by a changing magnetic field. Because the charge experiences a force that varies in time, the electric field from the charge varies in space. That is, a time varying magnetic field produces, in a real sense, a spatially varying electric field.

By the middle of the nineteenth century, the British physicist James C. Maxwell deduced, through a logical study of capacitor charging, that a current not associated with a real flow of charge, but rather with the rate of change of an electric field, existed. This current, called a *displacement current*, accounted for the deposition of charge on the plates of a capacitor. In addition, the displacement current demonstrated symmetry in electromagnetic phenomena that Maxwell recognized and was able to codify in a series of four equations. *Maxwell's Equations* hold the same significance in the study of electromagnetism as *Newton's Laws* does in the study of mechanics.

From the time Maxwell published his laws until the last decade of the nineteenth century is often thought of as the high point of classical physics. For thirty years, physicists were fairly sure all fundamental physical laws had been discovered and the science would soon become one of filling in details and determining fundamental numerical constants to greater degrees of accuracy. As we shall discover in later chapters, the classical physicists could not have been more wrong.

QUESTIONS

1. What are the two schools of thought concerning the forces between bodies?
2. What is the definition of a force field?
3. Is a force field a vector or a scalar quantity? Why?
4. What is the quantitative definition of the electric field?
5. How is the gravitational field defined?
6. What does the shape of a magnetic field produced by a bar magnet tell us about the existence of magnetic monopoles?
7. What did Biot and Savart contribute to our understanding about the source of magnetic fields?
8. How is the ampere determined?
9. How is the motion of a charge particle affected when it moves into a region of magnetic field?
10. How was Faraday able to demonstrate the existence of an induced current?
11. How did Henry's experiment differ from Faraday's?
12. What is the average value of an alternating current?
13. Why must alternating current be used with a transformer?
14. How does a step-up transformer differ from a step-down transformer?
15. What is Lenz's Law?
16. How does "displacement current" differ from "conduction current"?
17. What did Maxwell contribute to our knowledge of electromagnetic theory?
18. What period of time is considered the pinnacle of classical physics?

CHAPTER IX

LIGHT AND ELECTROMAGNETIC WAVES

Because most of the information that humans process enters the brain via the eye, it is logical to assume that light, the medium for the transmission of visual information has always been of great importance and interest. The ancients thought of light as a stream of particles emanating from a bright object and detected by the eye. This model of light formed the basis of Newton's theory of the nature of light published in 1675. However, the particulate theory of light did not explain some of its observed phenomena. Christiaan Huygens proposed an alternative theory of light in 1678. Huygens' hypothesis was that light was a wave somewhat analogous to a water wave. He used his wave hypothesis to explain the phenomenon of polarization in 1679. By 1690, he proposed what we now know as *Huygens' Principle: Every point on an advancing light wave acts as a secondary source of waves.* This principle was used quite successfully during the eighteenth and early nineteenth centuries to explain interference and diffraction phenomena. By the middle of the nineteenth century, Maxwell had successfully derived a theoretical argument that light consisted of electromagnetic waves that propagated through space with a measurable constant velocity. In 1873, he completed his study of the electromagnetic nature of light and predicted the existence of other electromagnetic waves beyond the sensitivity of the eye. His predictions of such waves, now know as radio waves or microwaves, was demonstrated in the laboratory by Heinrich Hertz in 1887.

The triumph of the electromagnetic wave theory of light seemed to be complete. In fact, the nineteenth century physics community believed there was nothing of fundamental importance to be discovered and that physics was a complete science, needing only more accurate measurements of certain universal constants of nature. By the middle of the first decade of the twentieth century, this belief was shattered. Great fundamental discoveries on the nature of light and matter were found during the decade from 1895 to 1905 that profoundly altered our concepts of physics. That the speed of light measured by any observer regardless of the observer's motion is constant is one of the fundamental discoveries leading to our modern understanding of nature.

The study of light's speed had fascinated observers since ancient time. Recorded attempts to measure the speed of light began with Galileo. His method consisted of placing two men with slide lanterns on mountains separated by approximately seven miles, as shown in Figure 1.

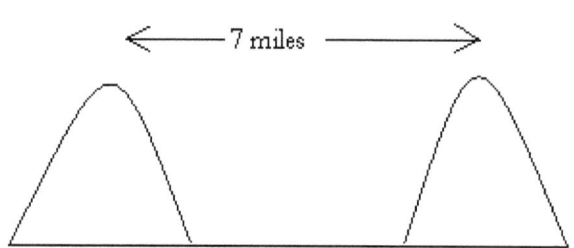

Figure 1.

The first man was to pull the slide of his lantern, which would be seen by the second man who would then pull the slide of his lantern. Galileo would measure the time between the first man pulling his slide and the second man's light returning to the first's position. By knowing the total distance traveled and the time of flight, Galileo would be able to calculate the speed. His conclusion was, if not infinite light's speed is extraordinarily rapid.

The first successful attempt to measure the speed of light was performed by the Danish astronomer, Olaf Roemer in 1676. He observed the eclipsing of Jupiter's moons over the period of several years. By noting the difference in times for the eclipses measured six months apart, he was able to measure the time for light to travel the diameter of the earth's orbit about the sun to be approximately 22 minutes. From his imperfect knowledge of the diameter of the orbit, he calculated the speed of light as 2.1×10^8 m/s.

FUNDAMENTAL CONCEPTS OF PHYSICS

The Nineteenth Century was the time of great effort to measure the speed of light. In 1849, Armand Fizeau used a toothed wheel device to measure the speed of light as 3.15×10^8 m/s. Jean Foucault replaced the toothed wheel with a rotating mirror device which was later refined by Albert Michelson, American's first Nobel Prize winning physicist, who determined the speed of light to be 2.997930×10^8 +/- 300 m/s. D. H. Rank, using a calculation of molecular constants determined by IR and microwave spectroscopy, measured the speed of light to +/- 100 m/s. The presently accepted value, which is *defined* as the true value, is 2.99792458×10^8 m/s.

In many instances, (e.g. reflection, image formation, etc.) the wave nature of light is of very little consequence. The reason for this is that the wavelength of light is exceedingly small compared with dimensions with which we are familiar. In other instances, (e.g. interference, diffraction, polarization) the wave nature is of great importance. When light's wave nature is not important we are in the realm of *geometric optics*, when the wave nature is vital for understanding particular phenomena, we call the study *wave optics*, or *physical optics*.

An imaginary line drawn perpendicular to the front of an advancing wave is called a *ray*. We can use geometric optics to investigate the behavior of a light ray when it is incident on an interface separating one medium from another. In general, when such a situation occurs, at least a portion of the incident light ray is reflected at the interface. The law governing such reflection has been known since ancient times. Formally, we can state the *Law of Reflection: The angle between a light ray incident onto a reflecting surface and a line drawn perpendicular to the surface always equals the angle between the perpendicular line and the reflected ray.* This law is illustrated in Figure 2.

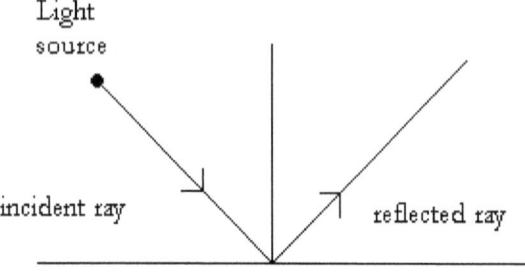

Figure 2.

In addition to the reflected ray, if the second medium is transparent, a portion of the initial ray travels into that medium. When entering the second medium, the ray's path is altered in a way illustrated in Figure 3.

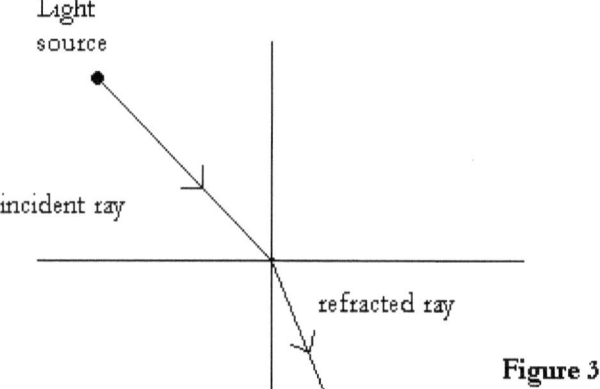

Figure 3.

The ray in the second medium is said to have experienced *refraction* and is called the *refracted ray*. The amount of bending that occurs when light passes from one medium to another depends on the properties of the two materials. In particular, *the ratio of the speed of light in vacuum to the speed of light in the material, called the index of refraction* of the material, determines the amount light is bent when going from one medium to the other.

FUNDAMENTAL CONCEPTS OF PHYSICS

While refraction had been observed since antiquity, Willebrod Snell successfully quantified the law of refraction in 1621. *Snell's Law states: The ratio of the sine of the angle the incident ray makes with the normal line to the sine of the angle the refracted ray makes with the normal line equals the reciprocal ratio of the indices of refraction of the incident and refractive material.* Snell's Law mathematical statement is given in Equation 1.

$$n_i \sin \theta_i = n_r \sin \theta_r \qquad \text{Equation 1.}$$

In this equation, n represents the index of refraction of the incident or refractive material, and θ is the angle between the ray and the normal line.

Example 1. What is the angle of refraction for light incident from air at an angle of incidence of 45^0 when it enters water that has an index of refraction of 1.33? The index of refraction of air at a pressure of one atmosphere is very close to 1, so, by Snell's Law the sine of the angle of refraction equals the sine of 45^0 divided by 1.33. Because the sine of 45^0 is 0.707, the sine of the refracted angle is 0.532. Thus, the angle of refraction is about 32^0.

The laws of reflection and refraction are useful in the study of image formation by optical elements. A *mirror* is a reflecting element, and a *lens* is a refracting element. It is possible to produce a myriad of optical instruments by choosing the shapes of the surfaces of lenses and mirrors, and using these elements in combination.

While knowledge of the direction of light rays is sufficient to obtain a great deal of information concerning image location and character, some of light's more interesting properties depend on its wave nature. When investigating these properties, it is useful to consider a *monochromatic light wave*. This is a light wave that has a single wavelength. In nature, all light waves consist of many wavelengths and a purely monochromatic light source is an idealization. In practice, however, lasers are extremely close to producing monochromatic light waves. However, lasers were not developed until the middle of the twentieth century. Before that time, some clever devices were used to produce approximately monochromatic light sources. Thomas Young, the British scientist, used one such device. He allowed the yellow light from burning sodium to fall onto a single narrow slit in an obstruction. The light from this single slit then fell onto two other closely spaced slits in another obstruction. These two slits acted as two sources of monochromatic light waves. In 1802, Young described the interference of the light from the two sources.

Young's device produced two light sources that were not only monochromatic, but also, *coherent*. Coherent sources emit light waves that have a constant phase difference when they begin their travel. Young's experiment is illustrated in Figure 4.

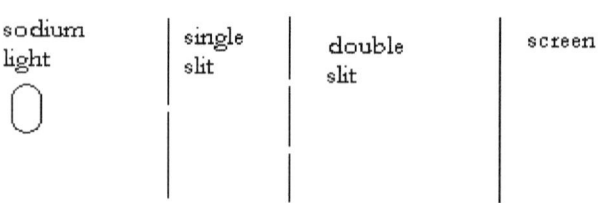

Figure 4.

Each point on the screen is a specific distance away from the double slits. At a point directly across from the double slits, the distance the light travels from the upper and the lower slit is exactly the same. Therefore, the light waves from the two slits arrive at this point on the screen with the same phase. As with any waves, when two waves arrive at some point in phase, their amplitudes add and we see *constructive interference*. At a point on the screen where the distances from the upper slit to the screen, and from the lower slit to the screen differ by half a wavelength, or any odd multiple of half a wavelength, the waves are completely out of phase when they combine. In that case, we see *destructive interference*. This situation is illustrated in Figure 5.

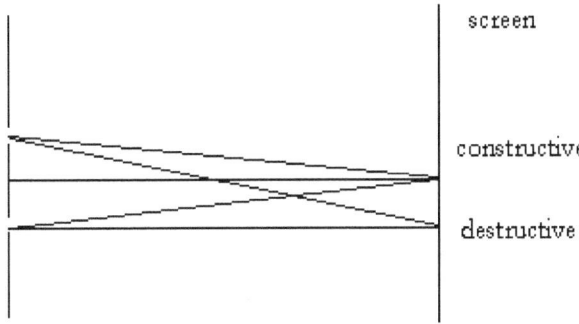

Figure 5.

Because visible light consists of a mixture of wavelengths from about 400 nm to 700 nm, respectively corresponding to the violet and red portions of the spectrum, different color light would produce a differently spaced interference pattern on the screen.

This dependence of interference patterns on the wavelengths of the light producing the pattern can be seen in the light reflected from a film of oil on the surface of a water pool. Whenever you see such a film, you notice it appears to consist of several colors. The reason is that light is reflected off the top surface of the oil and off its bottom surface. The light rays reflected at the two surfaces recombine in reflection, as shown in Figure 6.

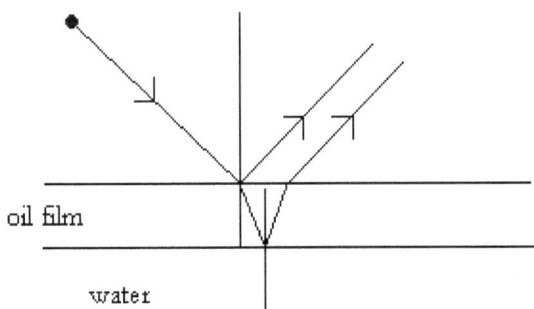

Figure 6.

Depending on the particular thickness of the oil film, these light rays may be in phase or out of phase. If they are in phase, there will be constructive interference on reflection. If they are out of phase, there will be destructive interference on reflection. This same principle applied to anti-reflective films that are used on lenses. The thickness of the film is chosen so the middle of the visible spectrum, about a wavelength of 550 nm, will suffer destructive interference on reflection. The extremes of the spectrum will not experience destructive interference so will be partially reflected. This is the reason anti-reflective coated optical elements are purple in color. Some red light is reflected as is some violet light making the lens look purple.

Determination of the wavelength content of light has been important in obtaining knowledge of the atomic and molecular structure. While it is possible to use Young's double slit as a device for observing the wavelengths present in light from a polychromatic source, the angular spread of the constructive interference lines makes this difficult if more than a very few wavelengths are present. To overcome this difficulty, we simply increase the number of slits maintaining the same spacing between adjacent slits. A device consisting of many slits is called a *grating* and can be used in a *spectrograph* to investigate the wavelength content of a polychromatic light source to a high degree of accuracy.

Now consider the situation where light from a monochromatic point source falls on a knife-edge obstruction such that a portion of the wave expanding from the source falls on a screen and part is obstructed, as shown in Figure 7.

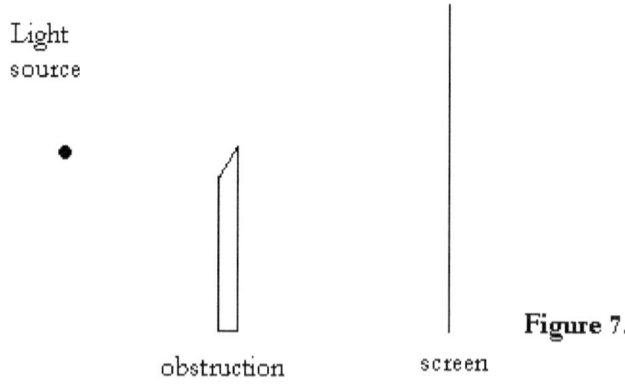

Light
source

obstruction   screen   **Figure 7.**

Instead of forming a sharp geometric shadow on the screen, the pattern of light on the screen consists of a shadow region, a light region, and a region between the dark and the light that consists of a series of dark and light bands. This band of fringes is called a *diffraction pattern.* *Diffraction* occurs whenever a portion of an advancing wave is limited by some obstruction. Because light waves are always limited by obstructions, diffraction effects are always present. However, because most sources are extended rather than point sources, the effects are usually masked.

Consider monochromatic light incident on a single slit as shown in Figure 8. Light from each portion of the slit travels a different distance in arriving at an observation point on the screen. Because of the infinite number of points on the line between the edges of the slit, it acts as an infinite number of separate interfering point sources of light.

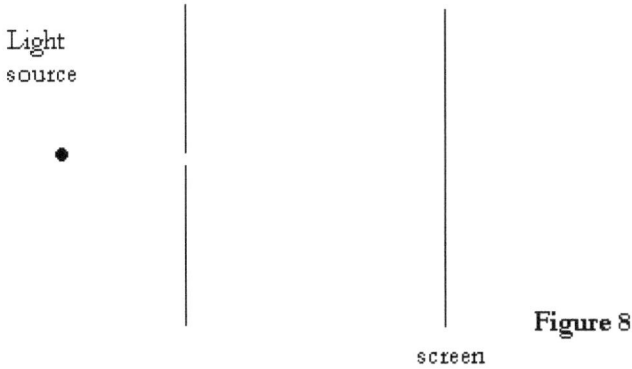

Light source

•

screen

Figure 8.

The pattern of light on the screen will have an intensity distribution where most of the light will have an angular spread determined by the wavelength of the light and the width of the single slit. The narrower the slit, the greater will be the angular spread of the light. On either side of this distribution will be dark bands followed by secondary light maximums. If the aperture is rectangular in shape, the diffraction pattern will spread out both vertically and horizontally depending on the height and width of the aperture. If the aperture is circular, a round diffraction pattern will be formed. As the diameter of the circular aperture decreases, the angular spread of the pattern increases.

Because every optical element acts as a limiting aperture, even the finest optical instruments are limited in their image forming ability due to diffraction effects. The very finest quality optical instruments are thus *diffraction limited.*

Consider two point monochromatic sources separated by some distance as shown in Figure 9. Light from each source is imaged as a diffraction pattern on the screen.

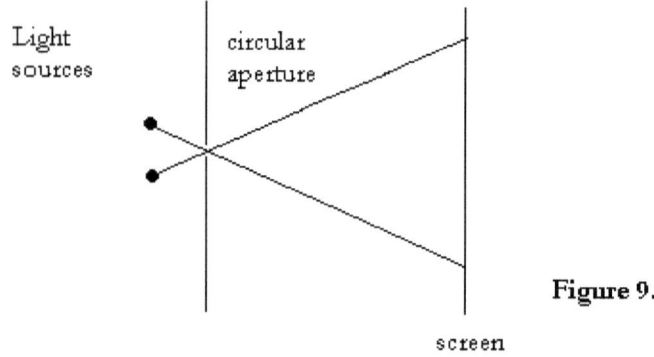

Figure 9.

Since each diffraction pattern is spread over the screen, there is a limit to the distance between the sources that can produce diffraction patterns that we see as distinctly resolved. Lord Rayleigh, the nineteenth and early twentieth century British physicist suggested the generally used criterion for resolution. *Rayleigh's Criterion states two sources are resolved when the minimum of the diffraction pattern from one source falls at the position of the maximum of the diffraction pattern of the other source.*

QUESTIONS

1. What did the ancients think of the nature of light?
2. How did Christiaan Huygens' thoughts of the nature of light differ from Newton's?
3. What is Huygens' principle?
4. What was Maxwell's contribution to the debate over light's nature?
5. What did Heinrich Hertz do to increase confidence in Maxwell's theory?
6. What was Galileo's conclusion concerning the speed of light?
7. What did Roemer observe in order to determine light's speed?
8. Why can light's wave nature be neglected when discussing image formation?
9. What is the difference between geometric optics and physical optics?
10. What is the Law of Reflection?
11. What is the index of refraction of a transparent material?
12. What is Snell's Law?
13. What is meant by the term "monochromatic"?
14. What is meant by the term "coherent"?
15. Describe Young's single slit experiment.
16. What is the condition for constructive interference on the difference in path for light from the two slits in Young's experiment?
17. How can you explain the appearance of bands of color in the light reflected from an oil slick on a pool of water?
18. What do we mean by a "grating"?
19. What is a spectrograph?
20. What is diffraction?
21. What is Rayleigh's Criterion for the resolution of two point sources?

CHAPTER X

RELATIVITY

Until now we have been studying *classical* as opposed to *modern* physics. The classical era lasted until the last decade of the nineteenth century. During the ten years from 1895 to 1905, the entire realm of classical physics was overthrown. We can say that classical physics is concerned with large things that move slowly. By large, we mean in comparison to atomic dimensions and by slowly we mean compared with the speed of light. Relativity, which is the subject of this chapter, concerns things that are large and move at an appreciable fraction of the speed of light. As we will see, if an object moves at less than 10% the speed of light, the difference between the predictions of classical and relativistic investigations is less than 1%. For most of the phenomena we experience in daily life, a classical approach is sufficient to meet our practical needs. However, some of the results of relativity, such as the equivalence of mass and energy, are significantly important for our understanding of nature.

The classical, or *Galilean*, theory of relative motion is familiar to most people. Consider a boat that can travel at a certain velocity in still water. If this boat enters a river and proceeds upstream, it will be moving with respect to a point on the shore with a velocity that is the difference in its velocity and the river's velocity. If it moves downstream, it will move with respect to the point on shore with a velocity that is the sum of the two velocities. Someone on a raft floating on the river will see the boat travel past it with a speed different from that observed by a person on shore. We can illustrate this in Figure 1.

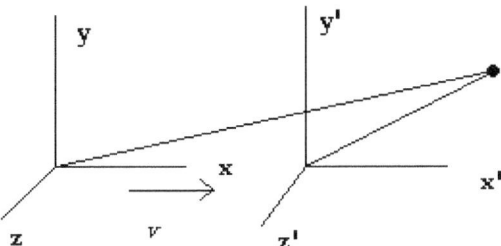

Figure 1.

An object's position is measured with respect to two coordinate systems. One system moves in the x direction with a speed v with respect to the other coordinate system. The coordinates of the object in the stationary system are (x,y,z) and in the moving system are (x',y',z'). The components of the object's velocity as measured by an observer in the stationary system are v_x, v_y, and v_z. In the moving system, the corresponding velocity components are $v_{x'}$, $v_{y'}$, and $v_{z'}$. If the two coordinate systems were coincident at some time that we can call t = 0, then the relationship between the velocity coordinates in the moving and stationary system are simply, $v_y = v_{y'}$, $v_z = v_{z'}$, and $v_x = v_{x'} + v$. These are the statements of Galilean relativity and are true as long as time is the same for observers in both coordinate systems.

That the laws of physics are the same for all observers moving with a constant velocity with respect to each other is one of our basic beliefs. The situation illustrated in Figure 1 is consistent with that belief in that force acting on the object will produce an acceleration that is the same for the observer in either coordinate system. However, a difficulty arises when we consider electromagnetic forces on a charged object.

Consider two charged objects moving with constant velocity with respect to a stationary observer as illustrated in Figure 2.

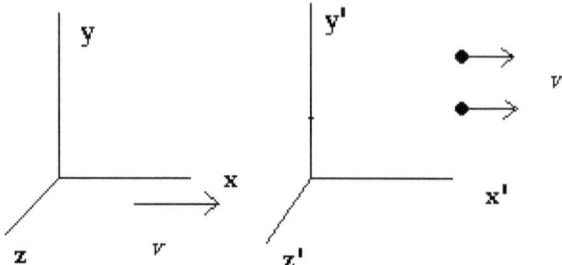

Figure 2.

The observer at rest in the unprimed coordinate system sees an electrostatic force between the charges and sees a magnetic force because the moving charges act as electric currents. An observer in the primed coordinate system, moving with the same velocity as the objects, does not see the objects as moving, thus observes no electric current and no magnetic interaction. This inconsistency, or lack of *invariance*, troubled theoretical physicists near the end of the nineteenth century and led to a fundamental rethinking of relativity theory that was formalized by Albert Einstein in 1905.

The lack of invariance of electromagnetic forces represents a failure of Galilean relativity. However, an earlier failure of Galilean relativity occurred in an experiment performed by the Americans Albert Michelson and Edward Morley in 1887. Michelson designed an interferometer to measure the change in the speed of light as a function of the orbital speed of the earth. Michelson believed, as did virtually the entire nineteenth century scientific community, light traveled as a wave through a medium called the "aether". Michelson pictured the passage of light through the aether much the same as the passage of a boat through water. He argued the orbital motion of the earth through the aether would produce an aether current similar to the water current in a river. The passage of light on the moving earth would be the same as the passage of a boat in a moving river of water. Michelson proposed to measure the aether current by means of an interferometer that would split a light beam into two, as shown in Figure 3.

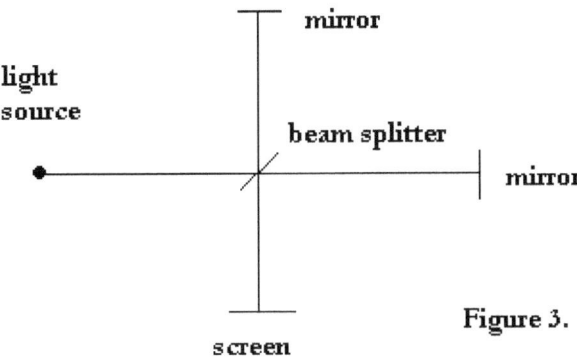

Figure 3.

One beam would travel in the direction of the aether current and the other would travel perpendicular to the current. The two beams would be recombined to form an interference pattern. The entire instrument was floated on a pool of mercury so that it could be rotated through ninety degrees. Now the beam that originally had been in the direction of the aether current would be perpendicular to it and the beam that originally was perpendicular to the aether current would be parallel to it. When the instrument was rotated, Michelson expected to observe a fringe shift of approximately one half fringe. In fact, he observed no fringe shift and considered his experiment a failure.

In an attempt to explain the null results of their experiment, Michelson and Morley speculated that the experiment was performed at the unlucky instant in time when the aether might be flowing with exactly the same velocity as the earth's orbital velocity. If this were the situation, performance of the experiment in six months should produce a maximum fringe shift because the relative speed of the earth with the aether would be twice the orbital velocity. When the experiment was repeated after six months, again there was no detectable fringe shift.

The experiment was performed throughout the year during different times of day and always produced the same null result. The scientific community was at a loss to explain the lack of a fringe shift.

In 1889, the Irish astronomer, George Fitzgerald, proposed the lack of fringe shift could be explained if the arm of the interferometer directed parallel to the orbital velocity of the earth were contracted such that its new length was related to its original

103

length by a factor of the square root of one minus the ratio of the square of earth's orbital velocity to the square of the velocity of light. This can be written in an equation as:

$$l = l_0 \sqrt{1 - v^2/c^2}$$ Equation 1.

This equation predicts the very interesting result that an object's length approaches zero as its speed approaches that of light. Because Fitzgerald's argument was very empirical, the scientific community was not very convinced of its validity.

However, Hendrick Lorentz, in 1895 derived Fitzgerald's contraction equation from more fundamental, theoretical arguments. Lorentz's arguments seemed somewhat strained and a good explanation of the null results of the Michelson-Morley experiment awaited Einstein's publication of the Special Theory of Relativity in 1905. Einstein based his theory on the work of Lorentz, Joseph Lamor, and Henri Poincare. It was Poincare whose question of the existence of absolute time and simultaneity in 1898 led Einstein to his Special Relativity theory.

Einstein's theory depends on only two assumptions. His First assumption is that the laws of physics are the same in all systems that are stationary or move with constant velocity. This assumption implies there is no way to determine if a system is, in fact, at rest. One coordinate system is as good as any other coordinate system, as long as they move at a constant velocity with respect to each other. The second assumption is that the speed of light is constant to any observer. The implication of this assumption is that time is not absolute, but differs in a coordinate system that is moving at constant speed with respect to another coordinate system.

To understand the last statement, consider the two coordinate systems that we illustrated in Figure 1 to be coincident at some instant. At that exact instant when the origins of the coordinate systems are the same point in space, a light pulse is produced at the coincident origins. An observer in the unprimed system would see an expanding sphere of light with radius equal to the speed of light multiplied by time measured from the instant the pulse was produced. An observer in the primed system would also see an expanding sphere of light with radius equal to the speed of light multiplied by time measured in the prime system. Because the speed of light is the same for all observers, the only way these two observers can both be correct in their observations is if time differs in the primed and unprimed system. The equation for the light sphere in the unprimed coordinate system is

$$x^2 + y^2 + z^2 - c^2t^2 = 0 \qquad \text{Equation 2.}$$

In the primed system, the sphere's equation is

$$x'^2 + y'^2 + z'^2 - c^2t'^2 = 0 \qquad \text{Equation 3.}$$

The speed of light, c, can be the same only if t does not equal t'.

This is the argument that leads to the statement that time is the fourth dimension. This statement is not accurate. In fact, the fourth dimension is the square root of the quantity minus c squared t squared, which is not a quantity that exists in the real number system. However, the mathematics of complex numbers allows us to determine the relationship between the three spatial dimensions and time in two coordinate systems moving with a constant velocity relative to each other. It turns out the equations linking the two coordinate systems are exactly those derived by Lorentz. These Lorentz transformations lead us to relativistic truths about length, time and simultaneity.

Consider a rod of a certain length that is measured by observers in two coordinate systems as shown in Figure 4.

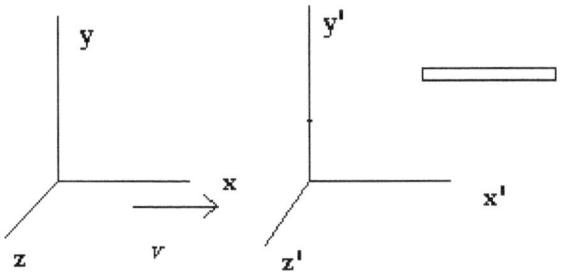

Figure 4.

The rod is parallel to the x (or x') axis and the primed system is moving with a speed v in the x direction with respect to the unprimed system. The measurement of the rod's length is simply found by taking the difference of its position coordinate in the x (or x') direction. Thus, an observer in the unprimed system measures the rod's length as $x_2 - x_1$ while an observer in the primed system measures $x_2' - x_1'$.

If the Lorentz transformations are used to relate the primed to the unprimed values, the length of the moving rod is less than a rod at rest by the exact amount required by Fitzgerald to account for the null result of the Michelson-Morley experiment. Moving lengths are contracted in the direction of their motion such that the moving length is related to its resting length by Fitzgerald's factor. This phenomenon of a reduction of a moving object's length is named the Lorentz-Fitzgerald contraction.

Example 1. What would we measure as the length of a meter stick traveling past us at 97% the speed of light?

Using Fitzgerald's length contraction equation, a stick with a resting length of one meter would have a contracted length given by L = 1meter times the square root of the quantity $[1 - (0.97c)^2/c^2]$. Doing the arithmetic leads to the conclusion we would observe the stick to have a length of slightly more than twenty-four centimeters.

We can use the same type of argument to consider the measurement of time extent in a primed and unprimed system. Because time transforms as a coordinate with a Lorentz transformation, we can surmise that time extent is just the difference in time coordinates in the same way length extent is the difference in position coordinates. Applying this reasoning we find that time runs slower in a moving coordinate system. Time itself is dilated.

Time extent in a moving system is related to time extent in a system at rest by the relationship

$$T_o = T\sqrt{1 - v^2/c^2} \qquad \text{Equation 4.}$$

T is the time extent in the resting system and T_0 is the time extent in the system moving with a speed, v. It is interesting to note that time on the object stops if the object moves at the speed of light. The fact that clocks keep different time depending on their motion leads to the conclusion that there is a lack of simultaneity in the measurement of events by observers who move with respect to each other.

Consider an observer in the unprimed system who drops a ball from a certain height and measures the time it takes for the ball to hit the floor. An observer in the primed system seeing the same falling ball will measure the ball falling the same distance in a different time. If both observers agree the acceleration due to gravity is constant, the moving observer must see the ball strike the ground at a different time than does the stationary observer. In other words, the fall of the ball does not occur simultaneously for both observers.

It is difficult for us to comprehend ideas of length contraction and time dilation because the effects are negligibly small unless we move very quickly. To help understand the reality of these phenomena, it is useful to review an experiment reported by D. H. Frisch and J. H. Smith in 1963. In their experiment, Frisch and Smith investigated the behavior of sub-atomic particles called muons created when primary cosmic rays enter the upper atmosphere. From terrestrial laboratory studies, we know muons are unstable and disintegrate in an average time of 2.2 millionths of a second. In this experiment muons with speeds of 0.9950 to 0.9954 that of light were detected near the peak of Mt. Washington at an altitude of 6265 ft. The detection apparatus was then moved to a position ten feet above sea level and adjusted to measure muons with the same range of speeds.

Because the distance traveled by the muons was about 1,900 m and their speeds were all about 3×10^8 m/s, classically it should take slightly over 6 millionths of a second to make the trip. But, again classically, they live only about 1/3 the needed time. Therefore, if classical physics is correct, there should be no muons detected at sea level. However, the experimental results show that most muons created at the altitude of 6265 ft. survived to reach sea level.

If we consider the muons' clocks are running slow as required by time dilation, a clock on a muon measures 2.2 microseconds in the same time that a clock on earth would measure 22 microseconds, certainly enough time to travel 1,900 m.

On the other hand, an observer on a muon would see the earth rising to meet the particle at a speed near the speed of light. The 1,900 m from mountaintop to seashore would be contracted to 189 m, certainly a sufficiently small distance for the particle to travel in its lifetime.

In the muon experiment, time dilation or length contraction represent observing a single phenomenon from different points of view. An observer on earth would say that time was dilated, while one on the muon would observe the length to sea level was contracted. This is true of all phenomena. Observers will see lengths contracted or clocks running slowly depending on their own frames of reference.

The relativistic requirement that the speed of light must be the same for all observers requires us to reexamine our concept of measuring velocities in moving and stationary coordinate systems. Consider again the situation illustrated in Figure 1. If the point whose coordinates we measured is moving, then an observer in the primed system and an observer in the unprimed system would measure different values for its velocity. For simplicity, let us consider the point is moving to the right. Because the primed system is also moving to the right with right with a velocity, v, if Galilean relativity is correct, an observer in the primed system would measure the point's velocity as:

$$v' = v - v \qquad \text{Equation 4.}$$

If the point were a light source, with light leaving the source with a speed, c, Galilean relativity would predict the observer in the primed system would measure the speed of light as c, while an observer in the unprimed system would measure the speed as $c + v$. This violates one of Einstein's two essential assumptions, and must be incorrect if the laws of physics are to be invariant.

To overcome this problem and obtain correct velocity transformations, we need only to consider that time is not the same in both systems. When we perform the required mathematical manipulations, we can obtain the correct relativistic representations of velocity. With these equations, every observer will measure, c, as the speed of light regardless of the observers' motion relative to each other.

Einstein published these conclusions in a 1905 paper. However, he did not end his study of relativity with these conclusions but considered the implications of relativity on dynamics. The most fundamental concept of classical dynamics is contained in Newton's second law of motion that states the net force acting on a particle equals the time rate of change of the particle's momentum. If no external force acts, momentum is unchanged.

Consider the elastic collision between two identical bodies as shown in Figure 5.

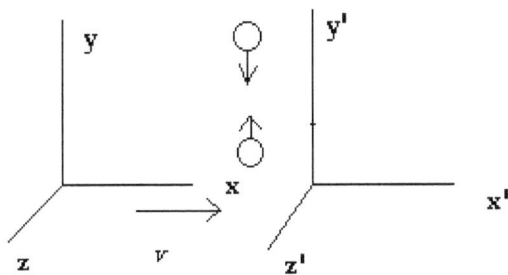

Figure 5.

In this figure, the top object moves only in the y direction as seen by an observer in the unprimed system and the bottom object moves only in the y' direction as seen by an observer in the primed system. The observer in the unprimed system sees the bottom object as having some velocity, v, in the positive x direction, while the observer in the primed system sees the top object as having a velocity, -v, in the x' direction. If no external force acts, each observer must measure no change in the overall momentum of the system of two particles. This is only true if the mass of a moving object is not constant, but is a function of velocity. This variation of mass with velocity is given by:

$$m_o = m \sqrt{1 - v^2/c^2} \qquad \text{Equation 5.}$$

The mass of a moving object is greater than it is when it is at rest, m_o. As predicted by Equation 5, when the object's speed approaches the speed of light, its mass approaches infinity. In order to accelerate the object any more, and infinite force would be needed. Thus, we can say it is impossible for any object that has a finite rest mass to travel at the speed of light. The speed of light is the absolute speed limit of the universe.

Example 2. What is the mass of an electron traveling at 0.95c?

The rest mass of the electron is approximately 9.1 x 10⁻³¹ kg. Therefore, it has a moving mass equal to this number divided by the square root of the quantity [1 − (0.95c)²/c²]. The result is 2.91 x 10⁻³⁰ kg.

Because an object's mass is a function of its velocity, we must define its momentum in a correct relativistic fashion as the product of its relativistic mass and its velocity. With this definition of momentum, Newton's second law of motion must consist of two terms. The force acting on an object must equal the product of the object's mass and its time rate of change of its velocity and the product of the object's velocity and the time rate of change of its mass.

When force acting on an object changes its kinetic energy, we now must consider the change in kinetic energy to be composed of two terms. The computation of an object's relativistic kinetic energy thus leads to the conclusion that its kinetic energy equals the difference between the product of its relativistic mass and the speed of light squared and the product of its rest mass and the speed of light squared, as stated in Equation 6.

$$E_k = \frac{m_o c^2}{\sqrt{1 - v^2/c^2}} - m_o c^2 \qquad \text{Equation 6.}$$

This equation states a free particle with no potential energy still has total energy that is the sum of the energy it has due to its motion and energy it has simply because of its mass. Another way of stating this fact is by noting mass and energy are convertible with the conversion factor being the square of the speed of light, $E = mc^2$. This huge conversion factor implies the release of a great deal of energy if mass can be changed into energy.

Example 3. How much energy would be produced if a one gram mass were completely transformed into pure energy?

Using the SI system of units, the mass is one thousandth of a kilogram and the speed of light is 3 x 10⁸ m/s. Squaring that speed and dividing by one thousand produces a figure of 9 x 10¹³ Joules, a truly immense amount of energy.

Although the success of Einstein's theory for non-accelerated systems was apparent shortly after its publication in 1905, Einstein himself considered it to be an intermediate result in the development of a theory that would apply to accelerated as well as non-accelerated systems. The 1905 theory is now called Einstein's *Special Theory of Relativity*, while one applicable to systems with acceleration is the *General Theory of Relativity*. Einstein produced his general theory after another ten years of effort, publishing his results in 1915. The central idea of the general theory is contained in the *Principle of Equivalence*, that states *inertial mass and gravitational mass are equivalent.*

We first introduced the concept of an object's mass as a measure of the amount of matter it contains. We considered the gravitational force between two massive objects, and we considered the force required to change the motion of a massive object. Until now there was no good reason to believe these two concepts of mass were actually concerned with the same thing.

If we examine the motion of a ball dropped near the surface of the earth, we measure its acceleration due to gravity to be 9.8 m/s^2. If we are in a spaceship drifting high above the atmosphere and release a ball we observe the ball float in space. If the spaceship's engines are fired such that the spaceship accelerates at 9.8 m/s^2, we will observe the ball fall with that acceleration. We cannot determine if we are near the earth's surface or are in an accelerating space ship simply by viewing the motion of the ball.

Consider now the situation where we direct a light beam through a small hole in an elevator car as shown in Figure 6.

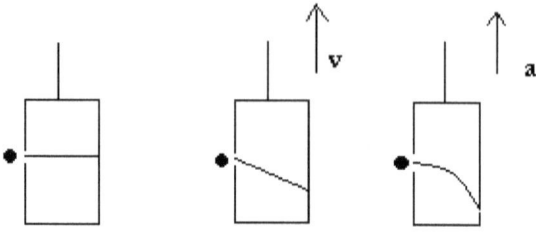

Figure 6.

The first elevator car is at rest and the light beam travels on the horizontal line to the opposite wall. The second car rises with constant velocity, v, and an observer on it sees the light beam travel the straight line to the opposite wall. An observer in the third car, accelerating upward with an acceleration, a, will see the light travel the curved path to the opposite wall.

As in the case of the dropped ball, the observer in the elevator car could not determine if his observations resulted because he was in an accelerated system or if he was in a gravitation field. As demanded by the Principle of Equivalence: *No observer in a closed laboratory can perform an experiment that distinguishes between effects produced by gravity or the acceleration of the laboratory.*

Again consider the situation in the accelerated elevator car. The light's path is curved, but light follows the shortest path between two points. Therefore, in the accelerated elevator car the shortest distance between the two walls is a curve called a geodesic. In the same manner, the shortest distance between two points in a gravitational field must be along a curved path where the curvature is greatest at positions in the field where the mass is greatest. Put another way, space is curved in the presence of massive bodies. This situation is illustrated in two dimensions in Figure 7.

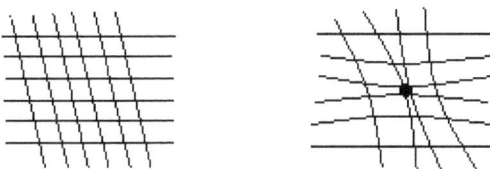

Figure 7.

In the first frame, there is no mass and the lines are straight and parallel. In the second instance, a massive object is in the center of the frame and the lines of the frame are curved inwards towards the mass. This is a result of general relativity that should be observable. Light beams traveling close to a massive body, such as a star, should travel on a path that is curved rather than straight.

In 1919, Arthur Eddington, the British astronomer and physicist, verified the bending of starlight during a solar eclipse. From his astronomical knowledge, Eddington knew a certain star would be behind the sun when it was in eclipse, as shown in Figure 8. If the star could not be seen during the eclipse it would mean that light traveling the straight line from the star to the earth would be stopped by the sun. If, on the other hand, the star were visible during the eclipse, Einstein's general relativistic prediction would be verified.

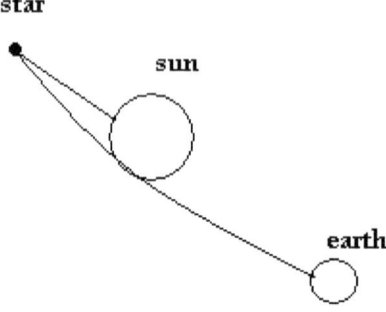

Figure 8.

Eddington observed the star, thus verifying Einstein's prediction.

One more triumph of general relativity is the explanation for the anomalous precession of Mercury's orbit. The elliptic orbit of this planet precesses about the sun at an observed rate of slightly over 5,600 seconds of arc per century. Classical Newtonian mechanics can account for almost 5,560 seconds per century, but no more. General relativity predicts an additional 43 seconds per century, bringing the theoretical predictions and the observed measurement into agreement within limits of observational error.

Another interesting and experimentally verifiable prediction of general relativity is the frequency shift of light traveling in a gravitational field. Light falling in a gravitational field is predicted to have its frequency shifted to higher values, while light leaving a gravitational field should have its frequency shifted to lower frequency or longer wavelengths. This gravitational frequency shift has been observed on numerous occasions providing another observed verification of general relativity.

QUESTIONS

1. When did the era of classical physics end?
2. What do we mean by "large" and "slow" when applied to classical physics?
3. How fast must an object move for relativistic considerations to produce a 1% effect?
4. How does the measurement of time differ in Galilean (classical) relativity and Einstein's special theory of relativity?
5. Why are mechanical forces invariant in Galilean relativity while electromagnetic forces are not?
6. Describe Michelson's interferometer.
7. How was the Michelson-Morley experiment performed?
8. What was Fitzgerald's explanation for the null results of the Michelson-Morley experiment?
9. What was Lorentz's contribution to the explanation of the Michelson-Morley experiment?
10. What assumptions did Einstein use to develop his special theory of relativity?
11. What is the "fourth dimension" as defined by special relativity?
12. What is meant by "time dilation" and "length contraction"?
13. Why are we able to say time dilation and length contraction are two different ways of looking at the same phenomenon?
14. Describe the Frisch-Smith muon experiment.
15. How does the Frisch-Smith experiment verify the predictions of special relativity?
16. Why can any object with a finite rest mass never reach the speed of light?
17. What is the relationship between mass and energy?
18. What is the difference between the special and general theories of relativity?
19. What is the "Principle of Equivalence"?
20. Describe three observations that support the predictions of general relativity?

CHAPTER XI

PARTICLE NATURE OF ELECTROMAGNETIC WAVES

Experiments with interference, diffraction, and polarization firmly establish the wave nature of light. However, when we attempt to explain the phenomena of light emission, absorption, and scattering, we are forced to conclude that light consists of a stream of particles that have no rest mass and are not fixed in number. These particles were first proposed by Max Planck in 1900 in order to explain the nature of the continuous spectrum of light emitted from a heated body. Although Planck originally proposed the quantum nature of light as a mathematical construct needed to force a theoretical equation to fit an experimental result, further experiments demonstrated the real existence of Planck's little bundles of light energy.

By 1900, experiments conducted on the emission of light from a small cavity within a heated object, sometimes called cavity or black-body radiation, had definitely established that the spectrum of Intensity (power per unit surface area) vs. Wavelength was continuous, with a maximum that varied with temperature such that the product of the peak wavelength and the absolute temperature, $\lambda_{max}T$, was constant, as shown In Figure 1 where T2 is greater than T1.

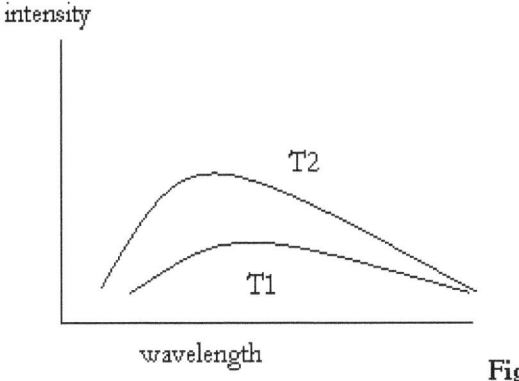

Figure 1.

Wilhelm Wien discovered the relationship between peak wavelength and temperature in 1893. He used this fact in his development of an empirical radiation law. The currently accepted value of the Wien constant is 2.9×10^{-3} m·K.

Wein's goal was to find a function of wavelength, $I(\lambda)$, such that the integration of $I(\lambda)$ over all wavelengths would produce experimentally observed result. Wein produced an empirical formula which approximated the experimental results in the short wavelength region but continued to rise to infinity as wavelength increased. Obviously, his attempt was not successful.

Example 1. From Wein's observation, calculate the temperature of the sun's surface. We know the sun's yellow appearance indicates it produces a radiation curve with a maximum wavelength of about 550 nm. Using Wein's law written symbolically as $\lambda_{max}T = 2.9 \times 10^{-3}$ m·K, we estimate the sun's surface temperature to be about 5,300 K. This estimate is somewhat cooler than the actual figure.

The next attempt to describe cavity radiation came from the British physicists Rayleigh and Jeans in 1900. In their statistical analysis of the problem, they assumed that dipole vibrators of atomic dimensions were excited in the cavity of the heated body and radiated energy in accordance with Maxwell's predictions. In allowing these vibrators any frequency of vibration, Rayleigh and Jeans had a continuous function that could be integrated over all wavelengths to produce the experimentally observed total intensity.

117

They were successful in predicting the experimental results in the long wavelength region of the spectrum, but again the total Intensity rose to infinity as the wavelength decreased.

Planck next took up the problem and followed the Rayleigh-Jeans approach but imposed the condition that the vibrators could take on only discrete values such that their energies could only have frequencies equal to hf, where h is a constant, now named for Planck, and f is the frequency of the vibrator. The equation stating this relationship is:

$$E = hf \hspace{3cm} \text{Equation 1.}$$

Because $I(\lambda)$ is a discrete rather than a continuous variable in Planck's derivation, the total Intensity is obtained by summing over all allowed frequencies rather than integrating as done by Rayleigh and Jeans. Planck's radiation law exactly fit the experimental observations. The Black Body problem had been solved, but a real question existed concerning Planck's theory. Were the little pieces of energy real or merely mathematical constructs that yield the correct prediction of the radiation law? Einstein gave the answer to this question in 1905, when he turned his attention to explaining the phenomenon called the *photoelectric effect*. He was awarded the Nobel Prize for his explanation of the photoelectric effect, rather than for his relativity theory.

In 1887, while performing the experiments to verify Maxwell's prediction that electromagnetic waves that propagate with the speed of light could be produced in the laboratory, Heinrich Hertz discovered that light incident on a metal surface could cause electrons to be emitted from the surface. In many respects this was not a surprising result. Light waves carried energy that could be absorbed by the metal surface and transferred to electrons in the metal in amounts sufficient to cause the electrons to be emitted from the surface. However, the wave picture of light would necessitate that the amount of light energy absorbed by the metal would be proportional to the intensity of the light shone on it. This could be experimentally evaluated by noting the maximum kinetic energy of the ejected electrons. If classical theory were correct, this energy should be proportional to the intensity of the light. Also, classical wave theory dictates the energy of a wave is related to its intensity, not its frequency. Thus, the maximum kinetic energy of ejected electrons should be independent of the frequency of light used in the experiment. In addition, if the light used in the experiment were of low intensity, there should be a time lag between the time the light is first shone on the surface and the time the first photoelectron is emitted.

All the predictions of the classical explanation of the photoelectric effect were demonstrated by experiment not to be valid. A new explanation was needed for this effect. Einstein, in 1905, proposed an explanation that fit with the experimental results. He proposed that light, in this experiment, behaved not as a wave but as a stream of massless particles, moving with the speed of light and possessing energy directly proportional to their frequency, as postulated by Planck. In this picture, the photoelectric effect is understood as a collision phenomenon where a quantum of light energy, now called a photon, collides with an electron bound to the metal. Because the maximum energy that can be transferred to the electron is the energy carried by the photon, there should be a linear relationship between the electron's energy, as measured by the potential needed to stop the electron from reaching the anode of the vacuum tube used in the experiment, and the frequency of the incident light. The equation describing this fact is:

$$eV_0 = hf - \phi \qquad\qquad \text{Equation 2.}$$

In this expression, eV_0 is the energy of the emitted electron, hf is the energy of the incident photon, and ϕ, called the work function, is the minimum energy that binds the electron to the metal.

Einstein's explanation gives a simple solution to the other puzzling aspects of the photoelectric effect. In his view, the intensity of the light is nothing more than a measure of the number of photons contained in the light beam. Therefore, the photocurrent in a particular experiment should be directly proportional to the number of photons incident on the surface as determined by the beam intensity. There should be no time lag between the first shining of the light on the surface and the emission of the first electron if, indeed, the phenomenon is a photon-electron collision.

Einstein's explanation of the photoelectric effect firmly established the photon picture of light and vindicated Planck's analysis of cavity radiation as being more than a mere mathematical trick.

Once the photon concept was established, it was used to explain other phenomena. Arthur H. Compton, in 1923, used the photon picture to explain the scattering of X-rays that are like visible light but of significantly higher energy. Because the energy of an X-ray photon is so much greater than the energy with which an electron is bound in an atom, Compton reasoned that the scattering of X-rays could be explained as a photon-free electron collision as opposed to the photoelectric effect that is a photon-bound electron collision. When considered in this way, the collision of the photon and the free electron had to occur with the conservation of momentum and total energy. Thus, Compton predicted the scattered X-ray photon would experience a frequency shift to lower frequency that depended on the scattering angle. His explanation perfectly matched the experimental observations. Thus, the photon picture was well accepted by the mid nineteen twenties.

It is interesting to note that X-rays, discovered by Wilhelm Roentgen in 1895, are produced in a manner that is just the opposite of the photoelectric effect. To produce X-rays, a high potential difference is placed across two metal plates contained in an evacuated glass envelope. Electrons emitted from the low potential plate, called the cathode, are accelerated through the tube and collide with the high potential plate, the anode. The quickly moving electrons are brought to a sudden stop. This sudden deceleration of the charged electrons is accompanied by the production of electromagnet waves of short wavelength, *bremsstrahlung* or *breaking radiation*. Because so many electrons strike the anode, the bremsstrahlung radiation forms a continuous X-ray spectrum.

In addition to the bremsstrahung radiation produced when energetic electrons strike metal targets, mono-energetic X-ray photons with wavelengths that are unique to the anode material are observed. These *characteristic X-rays* were difficult to explain until Niels Bohr proposed a quantum theory of the structure of the atom in 1913.

The success of Planck's photon theory for the nature of electromagnetic radiation was assured by the experimental observations of the photoelectric effect, Compton scattering, and Bohr's atomic theory. By the early 1920's this picture of light was universally accepted. However, the earlier wave theory of electromagnetic radiation still produced exact explanations for phenomena including interference, diffraction, and polarization. How then are we to combine these two theories?

One way is to consider the nature of light is, in fact, dual. That is, in some instances a wave explanation is most appropriate and in other instances a photon picture is better in providing an explanation. This ambiguity in explanation is a problem of our language and experience rather than a problem of light's intrinsic nature. We live in a classical world where most objects are relatively large and move relatively slowly. The world of photons is a world of the very small. In this realm classical physics is inadequate and quantum physics must be used.

QUESTIONS

1. What types of phenomena are best explained by the wave nature of light?
2. How does the spectrum of light emitted from higher temperature black bodies differ from that emitted by lower temperature bodies?
3. What was Wien's contribution to the study of black body radiation?
4. How did Rayleigh and Jeans account for the electromagnetic radiation emitted from a heated cavity?
5. How did Planck's explanation differ from that of Rayleigh and Jeans?
6. How is the energy of Planck's quantum of radiation related to its frequency?
7. Describe the photoelectric effect.
8. What results did classical electromagnetic wave theory predict concerning the photoelectric effect?
9. How did Einstein approach the photoelectric effect problem?
10. How does the energy of the electron emitted in the photoelectric effect depend on the energy of the incident photon?
11. How does the photoelectric effect differ from the Compton Effect?
12. How are X-rays produced?
13. What is bremsstahlung radiation?
14. What defines the character of "characteristic X-rays"?
15. What is meant by the dual nature of light?

CHAPTER XII

ATOMIC STRUCTURE

Sometime around 400 B.C., The Greek philosopher Democritus proposed that all things were composed of indivisible atoms. This theory appeared again in 1804 when John Dalton, the British chemist, introduced the concept of atomic weights. Dalton continued to pursue his study, publishing laws of partial pressure and multiple proportions that firmly established atomic theory as the basis of chemistry by the second decade of the nineteenth century. In 1897, J. J. Thomson established that cathode rays consisted of electrons that originally were constituents of atoms and determined the ratio of the electron's charge to mass. In order to account for the overall electric neutrality of the atom, Thomson proposed an atomic model called the "plum pudding" model. In this picture, the atom is considered to consist of a sphere of positive charge in which electrons are embedded, "like plums in a pudding."

While Thomson's model explained the electric neutrality of the atom, it left several properties unexplained. One such property is the characteristic discrete spectrum of light emitted from an excited atomic gas. By the latter part of the nineteenth century, the study of the line spectral structure of light emitted from atomic sources was well advanced. Because hydrogen is the lightest and simplest atom, its spectrum was of considerable interest and was extensively studied. In 1885, Johann Balmer, a Swiss teacher, empirically found an expression that predicted the wave number (reciprocal wavelength) of the visible lines in the hydrogen spectrum. His equation is:

$$1/\lambda = R(1/2^2 - 1/n^2) \qquad \text{Equation 1.}$$

In this equation, λ is the wavelength of the particular spectral line, R is an empirical constant, and n is an integer greater than 2.

In 1890, Johannes Rydburg generalized Balmer's equation to predict the existence of a number of spectral line series in the hydrogen spectrum. Rydburg's expression is

$$1/\lambda = R(1/m^2 - 1/n^2) \qquad \text{Equation 2.}$$

In this expression m is any integer and n is any integer greater than m. The Balmer series of spectral lines is particular for m = 2. If m = 1 a series of spectral lines in the ultraviolet region of the spectrum, called the Lyman series, is predicted. If m > 2, various infrared series of spectral lines are predicted. There is nothing in the Thomson atomic model that would account for these observed spectra.

In order to test Thomson's model, Ernest Rutherford proposed a scattering experiment where alpha particles, which are particles carrying a positive electric charge and having mass roughly four times the mass of hydrogen, would be fired at a thin metal foil. The scattered alpha particles were to be observed by a graduate student, Ernest Marsden, who was directed by Hans Geiger. Sitting in a darkened room for hours on end, Marsden counting the flashes that occurred when scattered alpha particles collided with scintillation screens. The experiment, shown in Figure 1, gave the unexpected results that some alpha particles were scattered through large angles. Thomson's model cannot account for these results, and thus had to be discarded.

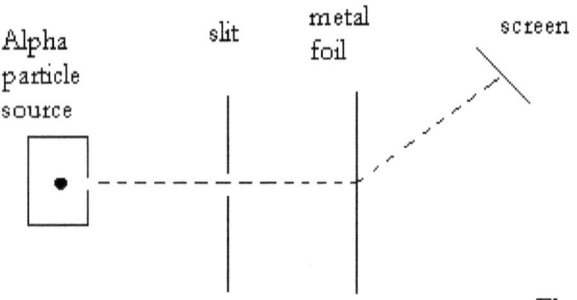

Figure 1.

FUNDAMENTAL CONCEPTS OF PHYSICS

When Rutherford attempted to make sense of the Geiger-Marsden results in 1911, he was forced to conclude that only very localized positively charged scattering centers could account for the large angle scattering. Therefore, he surmised the atom consisted of a small, dense positively charged nucleus with electrons revolving about the nucleus in a manner similar to the revolution of the planets about the sun. The electrostatic force of attraction between the positive nucleus and the negatively charged electron would bind the electron to the nucleus in the same way as the gravitational attraction between the sun and the planets keeps the planets bound to the sun.

In Rutherford's picture, alpha particles incident on the metal foil would in most instances pass through the atoms of the metal in the space between the electrons and the nucleus. A very few alpha particles would come close enough to a nucleus to have their paths affected by the repulsion between the nucleus and the particle. Still fewer would approach the nucleus head on and be backscattered. Rutherford made quantitative predictions on the fraction of alpha particles that would be scattered as a function of angle. His theoretical prediction fits the data of the Geiger-Marsden experiment to a very high degree of accuracy. It seemed Rutherford had an atomic model that explained the electric neutrality of the atom as well as the scattering experiment data. However, this model possessed one fundamental difficulty. The electron, being charged and moving in a circle thus undergoing angular acceleration, is an accelerated electric charge that according to Maxwell's classical electromagnetic theory must radiate energy and so spiral into the nucleus. Classical theory predicts the collapse of matter. Obviously, matter does not collapse. Neils Bohr in 1913 proposed his theory of the structure of the hydrogen atom that simply postulated atoms do not collapse when electrons are in certain stable orbits where classical laws do not hold.

Bohr was familiar with the Rydberg formula and with the results of the Geiger-Marsden experiment. His picture applied to hydrogen would be that of a single positive particle acting as the nucleus and a single negatively charged electron rotating about it as shown in Figure 2.

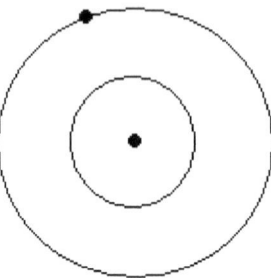

Figure 2.

In addition to stable orbits, Bohr postulated that the atom would radiate energy when the electron dropped from an orbit of higher energy to one of lower energy. The energy it radiated, in the form of an electromagnetic photon, would be the difference in energy between the electron's energy in the two orbits. To quantify the radii of the allowed orbits and thus the energy the electron possessed in any orbit, Bohr postulated the angular momentum of the electron must come in integral units of Planck's constant divided by 2π. While this seems to be an arbitrary requirement, the Bohr theory works in the sense that the theory predicts with great accuracy the wavelengths of the hydrogen spectrum.

If the stable orbits are numbered beginning with the smallest radius orbit, the energy of the electron is proportional to $1/n^2$ where n is the orbit number. The difference in energy between two orbits would be proportional to $1/n^2 - 1/m^2$, where n and m are the numbers of the orbits. This is very similar to the Rydberg formula predicting the wavelength of spectral lines. The theoretical prediction based on Bohr's theory and the experimental observations of Balmer and other spectroscopists were in very close agreement, leading to acceptance of the existence of stable energy states within the hydrogen atom. Emission spectral lines arise from the transitions of orbital electrons from orbits where they have higher energy to orbits of lower energy, as shown in Figure 3.

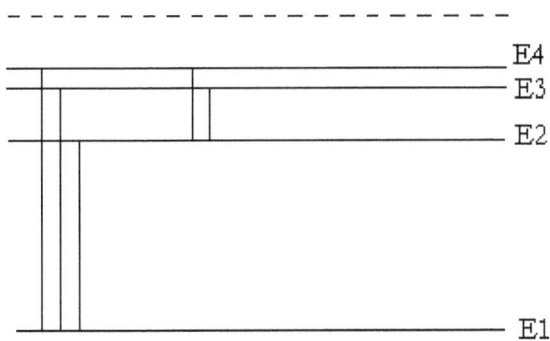

Figure 3.

The first set of transitions, where the electron ends on the E1 energy level, generates the Lyman series of spectral lines. The second set of transitions, ending on E2, produces the Balmer series of lines.

Bohr's theory marked a milestone in the development of our understanding of the atom. The essence of the theory defines the atom as a quantized energy system. This implies that atoms interact with the rest of nature in a quantized fashion. One example of this quantized nature is the emission of light quanta when atoms transition to lower energy states. Another example is the resonant absorption of photons by atoms causing atoms to rise from lower to higher energy states. This is seen in absorption spectroscopic experiments where light with a continuous distribution of wavelengths passes through a container of atoms. Certain wavelengths are preferentially absorbed. The absorbed photons have energies consistent with the energy level differences of the atoms.

The existence of line spectra is not the only indication that quantized energy levels exist within the atom. In 1914, James Franck and Gustov Hertz performed an experiment that demonstrated electrons lost energy in quantized amounts when they collided with atoms of a gas. In their experiment, Franck and Hertz used a glass tube filled with mercury vapor. A heated cathode in the tube provided electrons that were accelerated to the anode by a variable accelerating potential through a wire mesh grid. The grid was maintained at a potential slightly higher than the anode so that electrons with low kinetic energy will not reach the anode. Electrons that collide elastically with the mercury atoms will keep the vast

127

majority of their kinetic energy and will go through the grid to the anode, completing an electrical circuit. Electrons that collide inelastically will lose energy to the mercury atoms and will not have sufficient energy to reach the anode. Thus, when electrons lose energy on collision, the current measured by an ammeter will drop.

When they performed the experiment, Franck and Hertz observed a steady rise of current until a potential of 4.9 volts was reached. Then, the current dropped sharply but rose again with increasing applied potential until it reached 9.8 volts. Again, the current dropped but increasing potential caused it to rise until another drop occurred at 14.7 volts. Their results are shown in Figure 4.

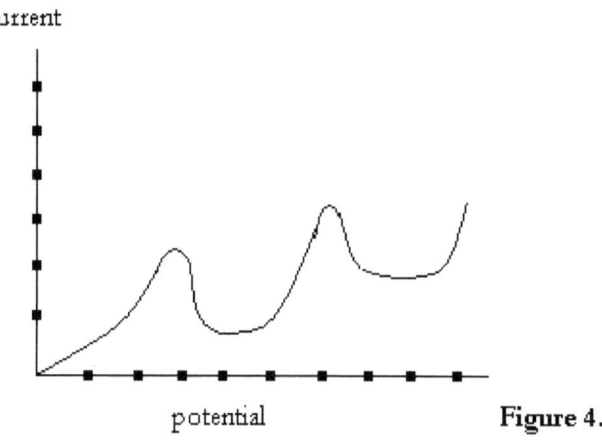

Figure 4.

Franck and Hertz interpreted their observed results as occurring because electrons with kinetic energy of 4.9 electron volts (eV) were able to excite mercury atoms from their ground (lowest energy) state to an excited state. The fact that the current broke in evenly spaced intervals indicated that electrons with kinetic energy greater than 4.9eV could resonantly collide with several atoms, giving up 4.9eV on each collision.

The mercury atoms that were excited would decay to the ground state with the emission of a photon having 4.9eV of energy. The wavelength of the emitted photon would be 235nm. The analysis of the light emitted by excited mercury vapor finds a spectral line with wavelength 235.6nm, demonstrating significant agreement with prediction.

The Franck-Hertz experiment can be modified so that the electrons can excite atoms to higher energy levels. Results of such modified experiments give a direct measure of the energy differences between allowed states. In all instances, wavelengths of light emitted from atoms excited by electron collision are consistent with observed spectral lines. Thus, the Franck-Hertz experiment directly verifies Bohr's hypothesis that quantized energy levels exist in atoms.

It seemed that the success of the Bohr theory and the Franck-Hertz experiment settled the concept of the atom as a mostly empty entity where electrons surrounded a nucleus in quantized energy states. While this concept is largely correct, it is incomplete. The Bohr theory fails to make any predictions concerning the relative intensities of spectral lines. It works best for hydrogen and hydrogen-like atoms such as singly ionized helium. However, for atoms as complicated as neutral helium, the theory failed. When the resolution of spectroscopic instruments improved, study of spectral lines revealed they were not monochromatic but consisted of a fine structure of several closely spaced wavelengths.

In 1916, Arnold Sommerfeld proposed that electrons actually revolved about the nucleus in orbits that were elliptic, as are the orbits of the planets in the solar system. Sommerfeld determined the energy of an electron in an elliptic orbit depended on the two axes of the ellipse. Because the energy now depends on two quantum numbers, there can be several states possessing the same total energy. Such states are called *degenerate*.

Sommerfeld noted that the relativistic variation of an electron's mass with its speed in an elliptic orbit would produce energy states with the same principal quantum number but slightly different values of total energy. When he applied these relativistic considerations, he discovered a fundamental constant of nature called the *fine structure constant,* α, that is related to other fundamental constants, the speed of light, the electronic charge, and Planck's constant. Numerically, α is very nearly equal to $1/137$.

The energy level diagram for a hydrogen-like atom now needs to be modified to show the substates within a given level as shown in Figure 5.

Figure 5.

The only problem with the Sommerfeld energy level scheme is that not all transitions between upper and lower states are observed. In the case of the two possible transitions from the first excited n = 2 state to the ground state, only one is observed. Likewise, of the six possible transitions from the second excited state, n =3, to the n = 2, first excited state, only three are actually observed.

The Sommerfeld extension of the Bohr theory marks the high point of what is called *old quantum mechanics.* It helped explain the existence of fine structure, but it still did not address the problem of relative intensities. Additionally, the quantum theory of Planck, Bohr, and Sommerfeld seems to be an add-on rather than a theory based on fundamental truths. A fundamental theory had to await the last half of the nineteen twenties when theorists developed *new quantum mechanics,* based on ideas proposed by Louis de Broglie in 1923.

QUESTIONS

1. What is the essence of the atomic theory proposed by Democritus?
2. What was John Dalton's contribution to atomic theory?
3. What discovery did J. J. Thomson make in 1897?
4. Describe Thomson's "plum pudding" atomic model.
5. What contribution did Johann Balmer make to explain the line spectrum of hydrogen?
6. How did Johannes Rydburg generalize Balmer's equation?
7. Describe the experiment Rutherford suggested to investigate atomic structure.
8. What were the results of the Geiger-Mardsen experiment? Why were they surprising?
9. How did Rutherford's model of the atom differ from Thomson's?
10. Why does Rutherford's atomic model violate the laws of classical physics?
11. How did Neils Bohr account for stable electronic orbits?
12. How did Bohr account for the emission of light from atoms?
13. Describe the Franck-Hertz experiment.
14. How did Franck and Hertz interpret their experimental results?
15. What were two failings of the Bohr theory?
16. How did Sommerfeld modify Bohr's theory?
17. What do we mean by "degenerate" energy states?
18. What fundamental constants are contained in the fine structure constant?
19. What is the numerical value of the fine structure constant?
20. What is a fundamental problem with the Sommerfeld theory?
21. Why is "old quantum mechanics" unsatisfactory?

CHAPTER XIII

WAVE NATURE OF PARTICLES

Louis de Broglie was a member of a prominent French diplomatic family and was expected to continue the family tradition by entering the field of politics and diplomacy. However, Louis had an older brother, Maurice, who had attained some success as an experimental physicist. Maurice completely supported the quantum theory of radiation and impressed Louis with the excitement of physics in the 1920's. Louis decided to abandon his study of history and turn instead to theoretical physics.

In his doctoral dissertation, completed in 1924, Louis de Broglie proposed that matter must have wave properties exactly analogous to the idea of the particle properties possessed by electromagnetic waves. This concept was so radical the faculty of science at the University of Paris needed the opinion of an outside expert to help it decide whether de Broglie should be granted his degree. The outside expert was Albert Einstein who saw that de Broglie's hypothesis was both valid and brilliant. Louis de Broglie received his degree and, after experimental verification of his hypothesis, a Nobel Prize in physics for his doctoral research.

The essential feature of his hypothesis is that the same relationships that connect mechanical and wave variables in Planck's quantum theory are valid for particles as well. Thus, there is a wavelength associated with a particle's momentum that is given by:

$$\lambda = h/mv \qquad \text{Equation 1.}$$

Also, there is a frequency associated with the particle's energy

$$\nu = E/h \qquad \text{Equation 2.}$$

The energy in the second equation is the particle's total energy that contains its relativistic rest energy.

Example 1. What is the de Broglie wavelength of a 1.0 kg shell with a muzzle velocity of 1,000 ms/? Using Equation 1 we divide Planck's constant, which in SI units has the value of 6.67 x 10⁻³⁴, by the product of the shell's mass and velocity. The result is a wavelength of 6.67 x 10⁻³¹ m. This is so small that it is impossible to detect.

Because h, Planck's constant, is so small, we would be unable to observe the wave characteristics of any reasonably sized object such as a baseball or a ballistic projectile.

While de Broglie's hypothesis is appealing in that it is complementary to Planck's theory, it is useless if it cannot be experimentally verified. This verification occurred in 1927 when two workers at Bell Laboratories in New Jersey, C. J. Davisson and L. H. Germer, unambiguously demonstrated electrons possessed a wave nature. They were investigating the scattering of electrons from a nickel surface with an apparatus pictured in Figure 1.

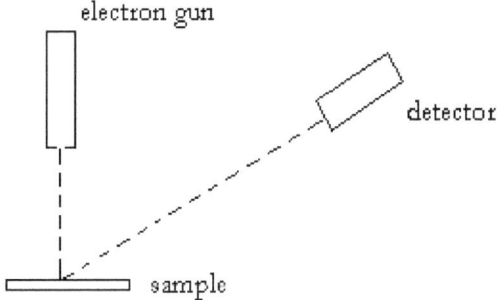

Figure 1.

Low energy electrons were shot from an electron gun onto a nickel sample. A detector recorded the electrons scattered from the nickel surface. A vacuum chamber surrounded the entire apparatus, and a clean surface was produced on the sample while it was in vacuum.

During the course of the experiment, the vacuum was broken and air entered the sample chamber contaminating the nickel. In order to save the sample, Davisson and Germer baked it in a high temperature oven to reduce the nickel oxide back to pure nickel.

When the sample was reinserted into the apparatus and the experiment resumed, the experimenters no longer detected a surface scattering pattern but rather a diffraction pattern.

In the course of heating and cooling the nickel sample, Davisson and Germer changed the sample's internal structure from polycrystalline to that of a single nickel crystal. When electrons fell on the single nickel crystal, they produced a diffraction pattern analogous to that produced when X-rays are incident onto a single sodium chloride crystal. By carefully measuring the angular position of the diffraction maxima and knowing the spacing between the crystal planes in nickel, Davisson and Germer were able to calculate the wavelength associated with the diffracted electrons. Their measurements agreed with de Broglie's theoretical predictions. This tended to verify de Broglie's rather remarkable hypothesis.

At the same time Davisson and Germer were diffracting electrons with a single crystal, G. P. Thomson, the son of J. J. Thomson who first measured the charge to mass ratio of the electron, was performing diffraction experiments with an electron beam incident onto a polycrystalline metal foil film. Because Thomson used higher energy electrons, the diffraction pattern he observed was sharper than that seen by Davisson and Germer. His experiment also agreed quite closely with de Broglie's predictions. These two experiments unequivocally demonstrated the wave nature of the electron. Davisson and Thomson shared the 1937 Noble prize.

Later experiments verified that all moving particles possess a wave nature. Louis de Broglie's assertion concerning the nature of matter was proven correct in all instances and necessitated a new approach in dealing with the motion of particles. Mechanics had to become wave mechanics.

The de Broglie concept of matter waves leads to a problem in the simultaneous measurement of some pairs of quantities. For any wave, the ability to simultaneously measure its position and the reciprocal of its wavelength is limited such that the product of the uncertainty of the two measurements is one. Likewise, if we want to simultaneously measure a wave's frequency and time is also limited so that the product of the uncertainties is one. In 1927, Werner Heisenberg considered the implications of these general statements about waves to the specific case of de Broglie matter waves. By relating a particle's momentum to the reciprocal of its wavelength, Heisenberg was able to write a limiting relationship stating *the product of the uncertainties of simultaneous measurement of a particle's position and momentum must be at least equal to Planck's constant divided by 2π.* This can be written as

$$\Delta x \Delta p_x \geq h/2\pi \qquad\qquad \text{Equation 3.}$$

This statement is known as *Heisenberg's Uncertainty Principle.*

Another statement of the principle concerns the attempt to simultaneously measure a particle's energy and time. In general, Heisenberg's principle represents a drastic departure from classical physics. The deterministic philosophy of Newtonian physics, that places no limits on the simultaneous measurement of any pairs of physical quantities, was incorrect. Nature herself places a limitation on the possible accuracy of measurement.

To get some feeling for the implications of Heisenberg's principle, consider the problem of locating the position of some particle. In order to see its position, we must shine some light onto the particle, but we can only measure the position of the particle to an accuracy of about one half the wavelength of the incident light. Therefore, to make a more accurate measurement we need to use a short wavelength light source. But, when looking at the source light from the photon picture, the beam of light consists of a stream of photons whose energy is reciprocally related to their wavelengths. As the wavelength decreases, the photon energy increases. From this point of view, high-energy photons are incident on the particle to be measured and collide with it. In the collision process, momentum is transferred from the photon to the particle, thus changing the particle's momentum. On the other hand, if we use a photon stream that doesn't disturb the particle's momentum, we will need low-energy photons of long wavelength. Thus, we will not be able to get a precise measurement of this position.

Until now, we have avoided the question of the nature of the matter wave. When we discussed photons as being small bundles of energy, we understood this energy was associated with an oscillating electric field. What is the analogous quantity that oscillates in the case of a particle?

To get an idea concerning this problem, consider again the intensity of a light beam as described by the photon picture and the wave picture. In the wave picture, the beam's intensity is proportional to the square of the of the electric field strength. In the photon picture, the intensity is proportional to the number of photons that cross an area perpendicular to the light beam per unit time, the *photon flux*. We can say, the photon flux is proportional to the square of the electric field. But, the photon flux is a measure of the probability of observing the light. Therefore, we are able to say the probability of seeing light is proportional to the square of the electric field wave function.

In 1926, Max Born argued an analogous picture must exist for matter waves. He held that in the same way the electric field wave function held within itself all the information needed to describe the photon, there must be a matter wave function that contained within itself all the information needed to describe a particle. He further held that as the square of the electric field wave function was proportional to the probability of locating a photon, the square of the matter wave function was proportional to the probability of locating the particle. By convention, the particle's wave function is usually designated by the Greek letter, ψ. Thus, we can say the probability of locating a particle is proportional to ψ^2.

This probabilistic interpretation of matter waves is called the Copenhagen interpretation and is the accepted norm. Although now accepted, the Copenhagen probability picture was not unanimously received. Einstein never accepted this view of physical reality. He maintained a deterministic viewpoint throughout his life.

Once Born proposed his probabilistic interpretation, it seemed reasonable that a problem would be solved once the wave function for that problem was determined. The last half of the 1920's and the first few years of the 1930's saw a great deal of work done on finding wave functions for many interesting problems. It is at this time that wave mechanics, or *new quantum mechanics*, came into being. On a very fundamental level, quantum mechanics is the basic truth of atomic and molecular physics and is the basis of all chemistry.

As we saw in the previous chapter, Sommerfeld's extension of Bohr's theory of atomic structure was unsatisfactory. It gave no explanation for unobserved transitions except to state those transitions were "forbidden". In addition, it made no statement concerning the intensities of spectral lines. A fundamental theory of atomic structure would need to explain these failings of the Sommerfeld theory. By the mid 1920's it was apparent that a fundamental reformulation of the laws of nature on the atomic level was needed.

FUNDAMENTAL CONCEPTS OF PHYSICS

The publication of de Broglie's hypothesis gave a clue to the direction researchers could follow in searching for this reformulation. When the circumference of the Bohr orbits was compared with the de Broglie wavelength of the electron in that orbit, it became apparent that only integer numbers of wavelengths were observed. This integer number requirement indicated to the Austrian physicist, Erwin Schroedinger, that an understanding of wave mechanics was required to solve the problem of atomic structure. In 1926, Schroedinger published the founding papers on wave mechanics and helped to further the revolution in thought from classical to quantum physics. In the following year, Heitler and London showed how quantum theory was able to explain chemical bonding.

At the same time Schroedinger was developing wave mechanics, Heisenberg led a group of theoretical physicists in the development of matrix mechanics. The results of wave mechanics and matrix mechanics are identical. It is only the mathematical formulation that differs in the two avenues of investigation.

The quantum mechanics that was developed by 1930 dealt with objects of atomic dimensions that moved at speeds that were appreciably less than that of light. P. A. M. Dirac, the British physicist extended the quantum mechanical investigation into the realm where speeds required relativistic considerations. In the early 1930's he published a series of investigations that established the discipline of *relativistic quantum mechanics*. With his work, we are now able to categorize our understanding of nature into four great regions differentiated by particle size and speed as shown in Figure 2.

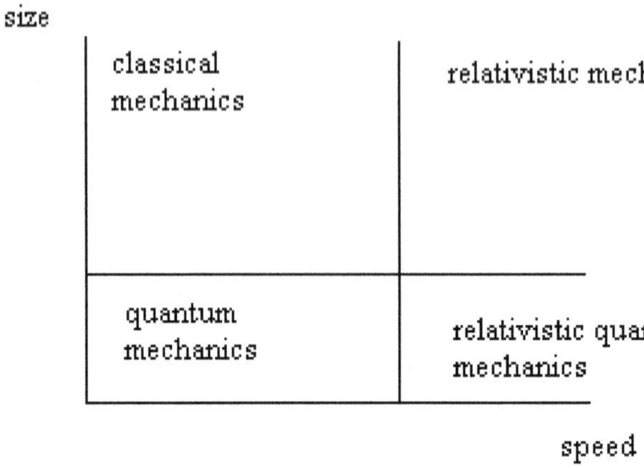

size

classical mechanics	relativistic mechanics
quantum mechanics	relativistic quantum mechanics

speed

Figure 2.

QUESTIONS

1. What is de Broglie's hypothesis?
2. Why do we not see the wave nature of matter in our normal observations?
3. Describe the Davisson-Germer experiment.
4. How did Thomson's experiment differ from Davisson and Germer's?
5. What is Heisenberg's Uncertainty Principle?
6. How should the wavelength of light be changed to better determine a particle's position?
7. What happens to the momentum of the light photon when its wavelength is shortened?
8. How did Max Born interpret the wave nature of matter?
9. When did the "new quantum mechanics" originate?
10. What was Schroedinger's contribution to the study of atomic structure?
11. How did Schroedinger's mechanics and Heisenberg's mechanics differ?
12. What contribution did P. A. M. Dirac make to our understanding of nature?
13. How does relativistic quantum mechanics differ from classical mechanics?

CHAPTER XIV

THE NUCLEUS

Rutherford explained the results of the Geiger-Marsden Experiment in terms of an atom consisting of a small, dense, positively charged nucleus surrounded by orbital electrons. At approximately the same time, J. J. Thomson was investigating the properties of positive atomic ions produced in discharge tubes. He determined the charge to mass ratio of several elements and discovered the greatest ratio for hydrogen, some 1836 times smaller than the same ratio he had measured for the electron. From this observation, Thomson surmised ionized hydrogen consisted of a single positively charged particle, the proton.

In other experiments, Thomson investigated elements such as neon that had an atomic mass of 20.2. Francis W. Aston, one of his students, showed that neon consisted of two isotopes with the same chemical characteristics but different masses. By the early 1920's we knew that nuclei of isotopes of the same chemical species contained the same number of protons but had differing masses.

To account for differing isotopic masses, investigators speculated some electrons could exist within the nucleus. The nuclear electrons would balance the charge of a number of protons in the nucleus leaving only the proper number of unbalanced protons needed to balance the charge of the orbital electrons of the neutral atom. Although attractive, this hypothesis was shown to be incorrect when Heisenberg's uncertainty principle was applied to the nucleus and demonstrated that an electron could not be confined to the nucleus.

The answer to the problem of isotopes was given by the British physicist, James Chadwick in 1932 when he discovered the existence of a neutral particle, roughly the same mass as the proton, within the nucleus. This particle is the neutron. Since that time our picture of the nucleus is of a very small, very dense, entity consisting of positively charged protons and neutral neutrons bound together by a force that is short ranged and significantly stronger than the Coulomb electrostatic repulsive force. Chadwick's 1932 experiment was a verification of a concept that he and Rutherford had believed since 1924. We now use the following notation to designate a particular nuclear species.

$$_Z^A X.$$

In this notation, Z, the atomic number, is the number of protons contained in the nucleus, A, the isotope number is the total number of protons and neutrons in the particular nucleus, and X is the symbol for the chemical species. Thus, the symbol for helium is

$$_2^4 He.$$

This notation shows the nucleus contains two protons and four total nucleons.

The diameter of the nucleus is about 100,000 times smaller than the diameter of the atom. Therefore, we get some idea that the strong force holding the nucleus together has a range of about 10^{-15} m. This force must be greater than the Coulomb force between protons for distance between them less than nuclear dimensions. A sketch of the potential energy between two nuclear protons is shown in Figure 1.

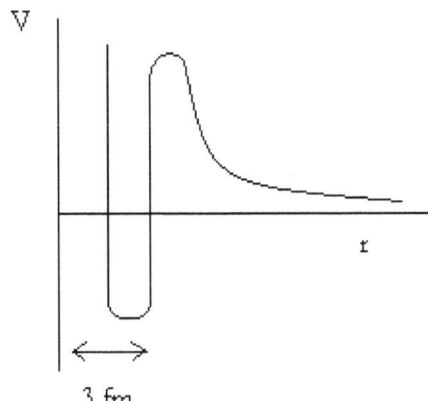

3 fm.

Figure 1.

When the distance between the protons is about 3 fm, that is 3 x 10^{-15} m, the strong nuclear attractive force overcomes the normal Coulomb repulsion between the particles. This force is approximately constant until the proton centers are very close. There appears to be a distance of closest approach between the protons. Therefore, we conclude there is some finite extent to the proton's size. This type of information concerning the forces between protons was obtained by experiments where beams of protons, ionized hydrogen atoms, were shot at proton rich targets. Such experiments are carried out in huge particle accelerators.

If neutrons are shot at proton rich targets to investigate the interaction between protons and neutrons, an energy function as shown in Figure 2 is obtained.

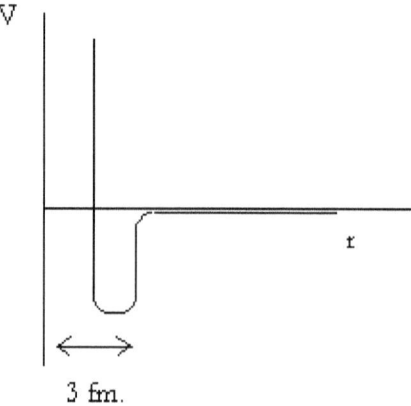

3 fm.

Figure 2.

Because the neutron is electrically neutral, there is no Coulomb repulsion between the proton and the neutron. Essentially no interaction occurs until the two particles are separated by about 3 fm. At this separation, an attraction very similar to that observed in the proton-proton interaction is seen. Again, as the distance between the particles becomes very small, extremely large repulsion is observed indicating the neutron, as well as the proton, has some extent in space. The strong nuclear force is not affected by the electric charge or lack of electric charge of the nuclear particles. The force between protons and neutrons is very strong and very short ranged.

The essential characteristics of the nuclear strong force was known by the Japanese physicist Hideki Yakawa, who in 1935 proposed this force was mediated by the exchange of a particle between the interacting nucleons. This particle is called the pi meson, or simply, pion. Yakawa's theoretically predicted particle was experimentally discovered in cosmic ray experiments in 1947 performed by Cecil Powell and his colleagues at the University of Bristol. Yakawa won the 1947 Nobel Prize and Powell the 1950 prize for their discoveries.

The concept of a particle mediating a force was useful in explaining the Coulomb force between electrons and nuclei in individual atoms. We can picture an exchange of photons taking place between the orbital electrons and their nuclei. Because the photon is massless, the Coulomb force is infinite in extent. Yakawa argued that the finite extent of the nuclear strong force required that the exchange particle for it must have a rest mass. He predicted the rest mass of the pion as approximately 200 times that of the electron. Powell's observations, verified Yakawa's prediction of the pion's mass.

Because nuclei are composed of charged particles and neutral particles that have some extent in space, we should not be surprised to learn that large stable nuclei contain more neutrons than protons. As the number of nucleons within the nucleus grows, the distance between nucleons can exceed the range of the strong nuclear force. When this occurs, the Coulomb repulsion between charged particles tends to favor the exclusion of protons from the nucleus. There are hundreds of stable nuclei that exist in nature. A plot of the neutron number of stable nuclei vs. proton number is shown in Figure 3. The line in this figure is the line of equal proton and neutron number.

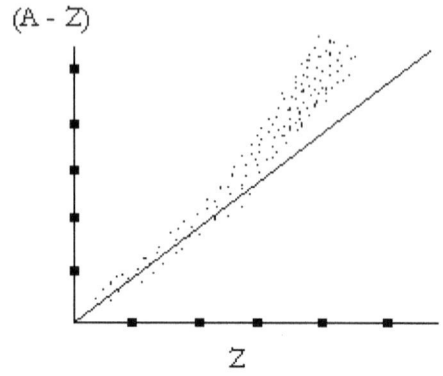

$(A - Z)$

Z

Figure 3.

Light stable nuclei tend to have equal numbers of protons and neutrons, while heavy nuclei are rich in neutrons.

The strength of the nuclear force is so great the total mass of a nucleus is less than the sum of the masses of its constituents. This mass difference represents the energy holding the nucleus together, its *binding energy*. We can determine the binding energy of a particular nuclear isotope by measuring its mass with a mass spectrometer and comparing it to the total mass of the protons and neutrons composing the isotope.

The simplest multi-particle nucleus we can investigate is the deuteron, the isotope of hydrogen consisting of one proton and one neutron. The deuteron is stable and has a mass we can accurately measure. When we compare its mass to the sum of the mass of a proton and neutron, we find the deuteron's mass difference is equivalent to energy of 2.225 MeV. We can demonstrate this is the energy holding the deuteron together by measuring the minimum energy of a photon that will disintegrate the nucleus when it collides with it. Such experiments show total agreement with theory. Therefore, we are confident we can determine nuclear binging energy simply by measuring nuclear mass

We are now able to consider the binding energy per nucleon for every nuclear species we might encounter. When we do this, we discover that the binding energy per nucleon basically increases with nucleon number until we have a nucleus consisting of roughly 80 nucleons. Nuclei with less than 80 nucleons can become more tightly bound if they fuse together while those with more than 80 nucleons can become more tightly bound if they split into smaller nuclei. Therefore, the process of nuclear fission of heavy nuclei or nuclear fusion of light nuclei releases energy. Fission energy is seen in nuclear power plants and nuclear bombs. Fusion energy is seen in the production of heat and light by the sun as well as the huge release of energy seen in a hydrogen bomb.

A plot of binding energy per nucleon vs. nucleon number is show in Figure 4.

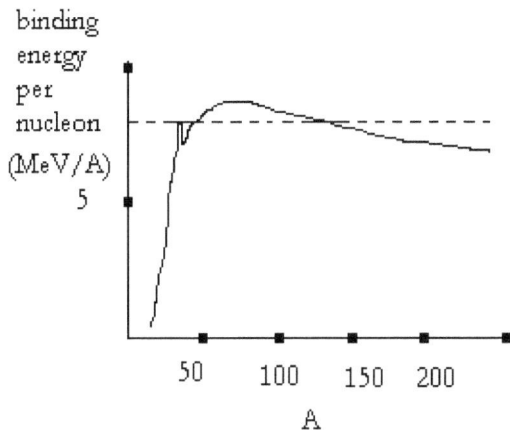

Figure 4.

The average binding energy per nucleon is approximately 8 MeV.

Shortly after Roentgen's discovery of X-rays, Antoine Becquerel, the French physicist, announced he discovered radiations from the element uranium that fogged photographic plates as did X-rays. Upon further investigation, Becquerel determined this radiation behaved as did X-rays in discharging charged bodies and that the rate of radiation production depended only upon the amount of uranium present in the sample.

Following Becquerel's discovery, Pierre and Marie Curie isolated two new elements, radium and polonium, that gave off radiations similar to uranium. It was the Curies who first called such elements *radioactive*. The Curies and Becquerel shared the 1903 Nobel Prize for their investigations into radioactivity.

Rutherford, in 1897, became interested in the newly discovered area of radioactive elements and found the radiation had two components. One component that was not very penetrating Rutherford called *alpha radiation*. The other component that could more easily penetrate matter he called *beta radiation*. By 1899, several investigators noted that beta radiation consisted of *negatively charged particles* that could be deflected by magnetic fields. Upon measuring the charge to mass ratio of these particles, they discovered beta particles were identical to the cathode rays J. J. Thomson investigated in 1897. We now know that both beta particles and the particles in cathode rays are electrons.

In 1903, Rutherford noted the deflection of a beam of alpha radiation in a magnetic field. The curvature of the particle beam indicated alpha particles were positively charged and much more massive than beta particles. An experiment performed by Rutherford and Ryods in 1909 determined the identity of the alpha particle as being identical to that of a *helium nucleus*, consisting of two protons and two neutrons. Elements that emitted alpha particles changed their chemical identity in the process.

In addition to alpha and beta radiation, Paul Villard discovered a very penetrating radiation in 1900. He named this radiation *gamma* radiation and noted it was unaffected by magnetic fields. In fact, gamma radiation was not particulate in nature but consisted of high-energy photons. Gamma radiation usually accompanies alpha and beta radiation.

We should note that alpha, beta, and gamma radiation arise due to changes taking place within the nucleus. Unstable nuclei can attempt to achieve a more stable configuration by ejecting alpha or beta particles. If a particular nucleus finds itself in an excited energy state, it can decay to a lower state by the emission of a gamma ray photon in a fashion analogous to the way an atom can eject a light photon when it drops from one excited state to a lower energy state.

Experimentally, it is relatively easy to determine the rate at which a sample of radioactive material ejects particles. This rate is called the *activity* of the material and is proportional to the number of unstable nuclei in the sample. We usually characterize a nuclear sample in terms of the time it takes for half the total number of nuclei in the sample to decay. This is called the *half-life* and can vary from very small fractions of seconds to a large number of years.

Consider a nucleus that ejects an alpha particle. The decaying nucleus, called the *parent* particle ejects an alpha particle consisting of two neutrons and two protons. After the decay, the original nucleus has changed into a different chemical species called a *daughter*. Alpha decay occurs in relatively large nuclei where two protons and two neutrons can come together to act as a unit. An example of such a reaction is the 212 isotope of bismuth that emits an alpha particle and transforms into the 208 isotope of thallium. We can determine the kinetic energies of the alpha particle and the daughter nucleus after the parent particle has decayed. This *disintegration energy* can only come from the energy associated with the difference in mass of the parent particle and the sum of the masses of the alpha and daughter particle. Therefore, we can show the kinetic energy of the ejected alpha particle is precisely fixed. George Gamow and the team of E. U. Condon and R. W. Gurney independently published the quantum mechanical theory of alpha decay in 1928.

The explanation of beta decay is a bit more involved than in the case of alpha decay. In the first place, the alpha particle is just a grouping of nucleons that already exist within the nucleus. However, in the case of beta decay, the electron does not exist as a free particle in the nucleus. Heisenberg's Uncertainty Principle prohibits a particle as small as the electron from being confined in a space as limiting as nuclear dimensions. Therefore, we must conclude that the nuclear electron to be expelled as a beta particle must be formed just before it is ejected. To account for the observation that the parent nucleus in beta decay change's chemical species but not isotope number, we are forced to conclude a neutron within the nucleus spontaneously changes into a proton and an electron that is ejected from the nucleus.

An example of a beta process is the decay of boron 12. After the decay, the daughter nucleus is carbon 12. The daughter has the same number of nucleons as does the parent, but possesses six protons and six neutrons while the parent has five protons and seven neutrons.

Beta decay cannot be explained on the basis of the strong nuclear force. There must be another *weak nuclear force* that is the cause of the decay process. If the process of beta decay were similar to that of alpha decay, one would expect to see beta particles ejected with discrete amounts of energy as are the alpha particles. However, the observed spectrum of beta particle energy appears as a continuous distribution shown in Figure 5.

number of beta
particles

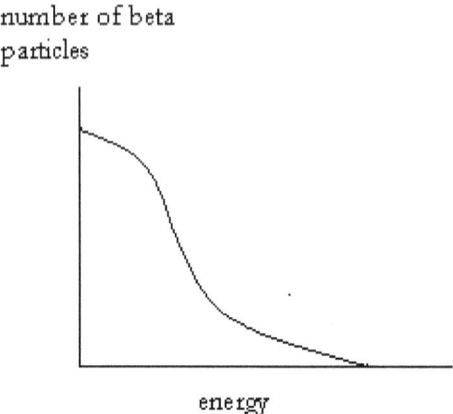

energy   **Figure 5.**

Behavior shown in Figure 5 seems to indicate a violation of the concept of energy conservation. Another experimental observation that the parent and daughter particles did not move in opposite directions after beta decay seems to indicate a violation of the conservation of linear momentum. Still another problem arises in the fact that the angular momentum of the parent does not equal the sum of the angular momenta of the daughter and the beta particle, showing an apparent violation of conservation of angular momentum.

In 1931 Wolfgang Pauli addressed these problems by proposing the process of beta decay involved the ejection of two particles from the parent nucleus, the beta electron and another very small, electrically neutral particle that would carry away with it some energy, linear momentum, and angular momentum. In 1934, Enrico Fermi developed Pauli's idea by proposing the beta decay process was initiated when a neutron within the nucleus disintegrated into a proton, an electron, and Pauli's little neutral particle (neutrino in Italian) Fermi's explanation has survived experimental test and is considered to be an accurate theory of beta decay. There is, however, a question as to the rest mass of the neutrino. In Fermi's theory, it has zero rest mass as does the photon. It may however, have a finite rest mass. Some of the most promising theories concerning elementary particles require the neutrino to have some rest mass. Currently, experiments show the rest mass of the neutrino must be less than the mass equivalent to energy of 10 eV. Because the neutrino is so small and carries no charge, it is extremely difficult to detect. However, very sensitive experiments have detected the existence of neutrinos, and have measured the rate of neutrinos arriving from the sun.

All nuclei more massive than bismuth 209 are unstable and decay by either alpha or beta emission. Upon study, it was obvious that the radioactive decay of heavy elements followed one of four possible patterns. These natural radioactivity series are named the thorium, neptunium, uranium, and actinium series. Because neptunium has a relatively short half-life of 2,250,000 years, natural examples of neptunium are not found.

QUESTIONS

1. How did Thomson conclude the nucleus of Hydrogen consisted of a single proton?
2. What is meant by the term "isotope"?
3. Why is it impossible for a free electron to exist within the nucleus?
4. What particle did Chadwick discover in 1932?
5. How does the existence of the neutron explain multiple isotopes of the same element?
6. How do we get an estimate of the range of the strong nuclear force?
7. How did Hideki Yakawa explain the nuclear strong force?
8. Why must Yakawa's pion possess a finite rest mass?
9. Why do heavy stable nuclei have more neutrons than protons?
10. Why does nuclear fission occur in heavy nuclei and fusion in lighter nuclei?
11. How do the particles in alpha radiation differ from those in beta radiation?
12. How does gamma radiation differ from alpha and beta radiation?
13. What is meant by the activity of a radioactive sample?
14. How does the distribution of energy of emitted alpha particles differ from that of emitted beta particles?
15. What does the continuous distribution of beta energies indicate about the mechanism of beta decay?
16. How is the apparent non-conservation of momentum in beta decay explained?
17. Why is the neutrino so difficult to detect?
18. Why is neptunium not found in nature?

CHAPTER XV

THE LARGE AND THE SMALL

The search for the ultimate constituent particles from which all matter is made was one of the most pressing quests of twentieth century physics. From Thomson's cathode ray experiments and Rutherford's alpha particle experiments to the particle scattering experiments carried out by the largest of particle accelerators, we have been attempting to identify the ultimate structure of matter. Although early information on elementary particles was obtained by observing cosmic ray events, progress in the field has been intimately associated with the development of particle accelerators where high-energy particles collide. Analysis of the fragments produced in such collisions is used to glean information concerning the structure of the colliding particles. After a great deal of effort, we now believe we understand the fundamental building blocks of matter and the ways in which they interact.

The other great area of investigation in the twentieth century was into the existence of the universe as a whole. Since Einstein's publication of his general relativity theory until the present time, our understanding of the nature of the universe has expanded. We are now very certain that the universe, as we know it, had a definite starting point in time some thirteen billion years ago and that it has expanded from a single point into the space it now occupies. We do not ask questions about time before the origin of the universe or of spatial extent before the universe expanded from a point. Such questions are beyond the realm of our scientific knowledge.

We are now confident that we have at least a general understanding of our universe and its fundamental constituent parts and interactions. This is similar to the position physics held at the end of the nineteenth century. As we have seen, that position was proven wrong by the advances of the twentieth century. Perhaps our current understanding will also be proven wrong. Perhaps later generations of physicists will view our current knowledge as an approximation to the truth that must be modified in light of new knowledge. Investigations in the twenty first century are needed to verify or modify our current understanding.

In our attempt to discover the basic building blocks of matter, we must first identify the basic forces that exist between particles. We knew of the gravitational force and the electromagnetic force before the start of the twentieth century. The nuclear strong and weak forces were discovered during the twentieth century, and now there is a hint of a fifth force that causes the rate of the universe's expansion to accelerate, sometimes referred to as "dark energy." While the existence of the fifth force is still speculative, we can characterize the four known forces in terms of their relative strengths. This characterization is shown below.

STRENGTH AND RANGE OF FUNDAMENTAL FORCES

FORCE	RELATIVE STRENGTH	RANGE
Strong	1	10^{-15}m
Electromagnetic	10^{-2}	∞
Weak	10^{-13}	$<10^{-15}$m
Gravitational	10^{-39}	∞

As early as the nineteen thirties, investigators were beginning to question the fundamental nature of the particles being discovered. We knew the radioactive emission of a beta particle was accompanied by the emission of a neutral particle of essentially no rest mass. The work of Paul A. M. Dirac on a relativistic quantum mechanics led to the conclusion that an anti-particle existed for every known particle. Antiparticles have the same mass but opposite charge and angular momentum as the known particles. Thus anti-electrons have the same mass but opposite charge than electrons. Because they are positively charged, anti-electrons are called *positrons*. C. D. Anderson observed the existence of positrons in the cosmic ray research he performed in 1932.

For some time, protons and neutrons were believed to be elementary particles, because they are the fundamental constituents of the nucleus. However, magnetic moment measurements indicated that protons and neutrons were themselves composed of constituent particles.

FUNDAMENTAL CONCEPTS OF PHYSICS

When Yukawa proposed his pion theory in 1936, many thought his predicted particle with a mass roughly 200 times the electronic mass was fundamental. Anderson's discovery of a particle with almost the same mass as that predicted by Yukawa seemed to verify Yukawa's theory. Anderson called the particle he discovered a mesotron, which was later shortened to *meson*. However, Anderson's meson did not have quite the characteristics predicted by Yukawa. We now call Anderson's particle a *mu meson*, or *muon*, to differentiate it from Yakawa's particle, the *pi meson* or *pion*. The pion was experimentally discovered in 1947. The mesons are not stable particles. They decay into other particles. Also, the mesons may possess either positive or negative electronic charge, or may be neutral.

The development of large particle accelerators in the nineteen fifties led to the discovery of vast numbers of particles, some with extremely short lifetimes. In order to attempt to characterize these "elementary particles", they are classified according to their rest masses. *Photons* have no rest mass. *Leptons* are particles with the smallest rest masses such as electrons and positrons. *Mesons* are particles with intermediate rest mass, and *Baryons*, such as the protons and neutrons have the highest rest masses.

This mass classification was useful for a time until discoveries were made that indicated particle characteristics were more important than particle masses. The muon, for example, has characteristics more like the electron than like the pion. Therefore, the muon should be classified as a lepton rather than as a meson.

A very fundamental classification can be made concerning the structure of particles, Electrons and muons, as well as neutrinos, seem to have no internal structure and are called *point particles*. Pions and baryons, on the other hand, seem to have an internal structure and are composed of constituent parts. Particles with an internal structure are called *hadrons*. Hadrons interact through strong, electromagnetic, and weak forces while leptons interact only through electromagnetic and weak forces. Strong forces are never a part of lepton interactions.

We now know that every lepton with a finite rest mass has a neutrino associated with it. Thus, there is an electron neutrino associated with the electron and a different muon neutrino associated with the muon. Additionally, as each particle has an associated anti-particle, each neutrino has an associated anti-neutrino. In fact, in normal beta decay, the ejected electron is accompanied by the electron anti-neutrino. In addition to the electron and the meson, a third type of lepton, called the tau lepton, has been discovered. This particle has a rest mass approximately 3,500 times that of the electron. The tau lepton has its own neutrino associated with it. Currently, we believe the three known massive leptons, their antiparticles, and their associated neutrinos form a complete set of leptons and that no more leptons remain to be discovered.

Distinct from the leptons that have no internal structure, the *hadrons* are elementary particles that possess an internal structure. This group of particles can be sub-classified into *bosons*, having integral values of spin angular momentum, and *fermions* that have half-integral values. The mesons are bosons while the baryons are fermions. The baryons may be further sub-classified into the proton and neutron and all other heavier baryons called *hyperons*. All the baryons other than the proton are unstable. The proton may be unstable with an extremely long lifetime on the order of 10^{30} years.

Hadrons differ from leptons in that they interact through strong or weak forces while the leptons do not interact through the strong force. In addition, hadrons have a spatial extent on the order of 10^{-15} m while the leptons are point particles. It seems that hadrons are not truly elementary particle but are composed of something more fundamental. Gell-Mann and Zweig formalized this concept in 1964 when they introduced the concept of a particle called the *quark*. Quarks are fermions that carry positive or negative electric charge equal in magnitude to $1/3$ or $2/3$ that of the electron. We now believe there are six such particle that, together with their anti-particles, form a complete set of truly elementary particle responsible for the structure of the hadrons.

Isolated quarks have never been detected and there is good reason to believe they never occur as single particles but always in combination. Two quarks combine to form mesons and three to form baryons. Current theory holds that the force between quarks is an aspect of the nuclear strong force but is mediated by particles called gluons rather than by pions.

We now believe there are six quarks that exist within three quark families. These six particles, together with the six leptons and their associated neutrinos, are now thought to be the ultimate building blocks of all matter.

FUNDAMENTAL CONCEPTS OF PHYSICS

We now turn to the problem of the universe itself. Until the early part of the twentieth century, physicists and astronomers considered the universe to be a static entity, neither expanding nor contracting. However, observational work performed by Edwin Hubble in the nineteen twenties indicated the universe was expanding rather than remaining constant in size. This expansion is the same in all directions. Everywhere we look, we observe galaxies receding from us with a speed directly proportional to their distance from us. The constant of proportionality is named *Hubble's constant*. The concept that all galaxies are moving away from all other galaxies implies the idea that there is no preferred direction in space and that the universe is, on average, the same in all directions. This *cosmological principle* forms the basis of our fundamental understanding of the universe.

The fact that all galaxies are receding from each other implies all the matter in the universe was contained in a single point of space at some time in the past. Some 13.7 billion years ago, the universe began to expand from this point. At that time a *Big Bang* occurred and the universe, as we now see it, came into existence. We are unable to say anything about time before the Big Bang because we have no idea what laws were operational before the instant of time marking the birth of our universe.

Until the nineteen sixties there was considerable debate in the scientific community about the Big Bang theory. Theoretical calculations were made indicating the universe cooled as it expanded until the present day when all of space would be filled with thermal radiation. This radiation must follow the Planck radiation law for a black body with a temperature of roughly 2.7 K. Arno Penzias and Robert Wilson made a most startling verification of the presence of such background radiation in 1965. While working on the problem of microwave communication with satellites, they discovered space to be entirely filled with microwave radiation having a peak intensity of 1.06 mm and following a Planck radiation law. Such radiation is exactly that predicted by the Big Bang theory. Further experiments have confirmed and refined Penzias' and Wilson's discovery to the extent that the entire astrophysical community now considers the Big Bang theory as a true representation of the state of the universe.

Recent experiments seem to indicate the expansion of the universe is accelerating. Additionally, recent observations lead to the belief that the matter we are able to observe is only a small fraction of the matter that actually exists. There seems to be unknown *dark matter* that is six times more abundant that the normal matter we can observe. Dark matter, together with normal matter makes up about 27% of the total mass-energy of the universe. The other 73% consists of *dark energy* which, although not directly observable, accounts for the acceleration of the universe.

QUESTIONS

1. How was information about elementary particles obtained before the invention of large accelerators?
2. What are the four know fundamental forces between particles?
3. What is the proposed "fifth force"?
4. What are the relative strengths of the four known forces?
5. What fundamental conclusion about the nature of elementary particles did Paul A. M. Dirac make?
6. What is a positron?
7. What contribution did C. D. Anderson make to our understanding of elementary particles?
8. Why did we conclude that protons and neutrons were not truly elementary particles?
9. Why was Anderson's discovery of an intermediate mass particle confused with Yukawa's predicted particle?
10. What is the difference between leptons and hadrons?
11. What is the difference between bosons and fermions?
12. How did Gell-Mann and Zweig account for the structure of hadrons?
13. What is a quark?
14. How many quarks are in a meson?
15. How many quarks are in a baryon?
16. What particles mediate the force between quarks?
17. What do we now consider to be the ultimate building blocks of matter?
18. What did Hubble discover concerning the size of the universe?
19. What is the Hubble constant?
20. What is the Big Bang?
21. What experimental evidence discovered by Penzias and Wilson solidified belief in the Big Bang?
22. Is the size of the universe increasing, decreasing, or remaining constant?
23. What is the source of the acceleration of the universes' expansion?

INDEX

www.ingramcontent.com/pod-product-compliance
Lightning Source LLC
Chambersburg PA
CBHW051520170526
45165CB00002B/543